LILIES
AND RELATED FLOWERS

P. J. Redouté

LILIES
AND RELATED FLOWERS

Text by
Brian Mathew

The Overlook Press
Woodstock, New York

First published in the United States in 1982 by The Overlook Press, Lewis Hollow Road, Woodstock, New York 12498

LIBRARY OF CONGRESS CATALOGING IN PUBLICATION DATA

Redouté, Pierre Joseph, 1759–1840
Lilies, and related flowers.

A selection of 108 plates from Redouté's Les Liliacées, published 1802–1816, with a new English text.
1. Lilies 2. Liliaceae I. Mathew, Brian
II. Stearn, William Thomas, 1911– III. Watson, William Patrick
IV. Title
QK495.L72R3213 655.9'3432 81-11021
ISBN 0-87951-135-4 AACR2

Filmset in England by Servis Filmsetting Limited, Manchester
Printed in Switzerland by Hallwag AG Bern
Bound in Switzerland by Maurice Busenhart, Lausanne

Frontispiece shows *Lilium superbum,* reproduced from a large-paper copy of *Les Liliacées*.

CONTENTS

The volume and plate numbers in parenthesis
refer to the placing of the illustrations in
the original eight-volume edition of *Les Liliacées*.

REDOUTE—ARTIST AND TECHNICIAN

William Patrick Watson

Pierre-Joseph Redouté, the Raphael of flowers, is unquestionably the botanical illustrator best known to the public, and his name has become synonymous with roses through his famous three-volume work which celebrates this lovely tribe. The decorative appeal of his original prints, and modern mass-reproduced copies of them, has popularized his name in a way unique among botanical artists. The achievement of Redouté is not so readily appreciated.

His genius is beyond dispute but he had his equals, comparatively unheralded, whose work can be found throughout the history of botanical art. Hans Weiditz who did the fine drawings for the woodcuts of the Brunfels herbal of 1531, and Albrecht Meyer who provided the illustrations for Fuchs' herbal of 1542, both display the artistic accomplishment of the age of Dürer in which they lived; in fact, much of Weiditz's work has at one time or another been attributed to Dürer. Nicholas Robert, in the late seventeenth century, produced a series of engravings of plants the magnificence of which has never been surpassed and, contemporary with Redouté, the Bauer brothers, Francis and Ferdinand, created some of the finest botanical illustrations ever.

Several other names immediately come to mind, so we must carefully define the reasons for Redouté's pre-eminence in an already illustrious field. The basis of this pre-eminence we will find to be three ingredients: a native genius, combining all the elements of diligent draughtsmanship of his predecessors, stubborn industry and unremitting toil; the political revolution of the Napoleonic Empire with its lavish patronage of the arts, which found personal expression in Redouté's role as protégé of Joséphine; and, accompanying these, a technical revolution in book production and illustration made possible by this patronage. And it is in *Les Liliacées*, Redouté's magnum opus, that all of these factors found their perfected expression.

Les Liliacées, eight folio volumes containing 486 plates, was published in a series of parts starting in 1802 and completed in 1816. Probably no more than 220 copies were produced; these in themselves represent a total of 10,920 separate plates! It is a much more varied work than its title suggests for, besides members of the lily family, one finds in it amaryllids, irises, orchids, and bromeliads, to name but a few. The illustrations were painstakingly colour-printed from engraved plates, and then 'finished' by hand. A few copies, no more than eighteen, are known to have received the personal attention of Redouté himself, adding highlights with a fine brush, and it is from such a copy that the illustrations in the present work are reproduced. But before discussing *Les Liliacées* we should look at Redouté's life, the mélange of events, accident and intent, that led up to this great work.

Pierre-Joseph Redouté was born in 1759 in the Belgian Ardennes. His father was a painter, as had been his grandfather and great-grandfather. He was the second of three sons, and all three were apprenticed in the craft of painting by their father at an early age; by the time he was thirteen Pierre-Joseph had sufficient skill and confidence to set out and seek employment. Of his wanderings little is known; his earliest recorded work is in 1775 when he assisted in the decorations of the Château de Carlsbourg in Bouillon. He undertook work for the Bishop of Liège, and then spent a period of time portrait painting in Luxembourg. For ten years he pursued an itinerant career as an artist 'interior decorator' as had his father before him. But his travels did bring him into contact with the work of the great Flemish masters and, in particular, that of the Dutch flower painters van Huysum and Ruysch. From remarks made later in his life we know that this exposure gave him the first indications of the métier he was to pursue.

In 1782, at the age of twenty-three, Redouté journeyed to Paris to join his elder brother, Antoine-Ferdinand, as one of the painters of stage scenery for the Théâtre Italien. During this period he began visiting the Jardin du Roi in his spare moments, drawing and painting flowers and plants. Some of these drawings were bought by the picture dealer Chéreau who had them engraved by de Marteau. They attracted the notice of Charles Louis L'Héritier de Brutelle, a wealthy amateur botanist who aspired to produce sumptuous botanical books illustrating some of the new species of exotic plants then being introduced into the gardens of Europe. L'Héritier persuaded Redouté to leave the theatre and work for him as an illustrator for his new book *Stirpes Novae*, for which he had already commissioned illustrations by Fossier, Prévost, and Fréret. The first of six parts appeared in 1784 and, although it did not contain any plates by Redouté, by the time the work ceased in 1789 (it was never completed as L'Héritier, by that time fallen into poverty, never regained his fortunes and was assassinated in 1800), fifty-four of the ninety-one published plates were by Redouté. In fact it seems that Pierre-Joseph, joined by his younger brother Henri-Joseph, took over production of the book, supervising its engraving, the printing of the plates, and the hand colouring of deluxe copies. Thus began a collaboration the effects of which were decisive to Redouté's career. L'Héritier instructed Redouté in the techniques of dissection and the details of plant anatomy, he supervised his drawings, making sure that the representations were botanically informative and correct, and he opened to Redouté his considerable private library, with its wealth of illustrated botanical books which Redouté could study and compare.

Assured of an income and with a generous patron, we might assume Redouté's life at this time to resemble an idyll, with the artist calmly and leisurely painting plants, detached from the turmoils of the day. But plants and politics were not unmixed: Joseph Dombey, physician and botanist, had been commissioned by the French and Spanish governments to collect new and useful plants in Chile and Peru and to bring back seeds and living and dried specimens. One set was to go to the Spanish government, another set to be retained by him. As events chanced the consignment destined for Spain was lost, and so Dombey's own collection was seized and divided, Dombey taking to Paris a miserable but important remnant of the vast herbarium he had assembled. L'Héritier, hearing of this, persuaded Buffon to entrust the herbarium to him, for here was a chance to publish a major work on an important and new flora, utilizing the talents of Redouté. The Spanish government however learned of this intended publication and protested to the

French authorities. L'Héritier, learning of the repossession order at the Court of Versailles, hurried home and, with Redouté, worked through the night packing the voluminous collection of dried plants. With the herbarium in his possession, he slipped away the next day to London, where he was received by Sir Joseph Banks in Soho Square. Here Redouté joined his patron in April of 1787. Along with the continuing work on the *Stirpes Novae*, L'Héritier commenced his *Sertum Anglicum*, a lovely book illustrating many of the plants recently introduced in the gardens of Kew, and intended as a homage to the hospitality of his English colleagues. This book, published in 1789, contains thirty-five plates, out of which twenty-two are by Redouté, most of the remainder being by another distinguished protégé of L'Héritier, James Sowerby. While in London Redouté met the Florentine engraver, Francesco Bartolozzi, whose influence on Redouté we will examine later.

L'Héritier and Redouté did not stay long in London. The beginning of 1788 saw them back in Paris, Dombey's herbarium safely reinstalled in L'Héritier's library. It was at this time that the great flower painter, Gerard van Spaendonck, Professeur de peinture de fleurs at the Jardin des Plantes, requested Redouté to contribute to the famous Collection des Velins. This collection of natural history paintings on vellum, begun in the mid-seventeenth century by Gaston d'Orléans, had passed into the King's Library and was regularly added to by the outstanding artists of the day. Nicholas Robert, already mentioned, had been one of its earliest contributors, and a comparison of his work with that of Redouté for the *Stirpes Novae* leaves little doubt that his influence was felt. But it was Redouté's close contact with van Spaendonck that was to prove decisive to his art. Van Spaendonck was a follower of van Huysum whose work had already had such an impact on Redouté during his early wanderings. Although van Spaendonck did not share Redouté's botanical enthusiasms, and instead treated flowers as pure naturalistic art forms, his technical finesse opened Redouté's eyes to new ways of depicting plants and more accurately presenting their delicate freshness and gradations of hue and tone. For his paintings on vellum for the collection, van Spaendonck abandoned the heavy, opaque medium of gouache, and used pure watercolour instead. He also produced an exquisite and now excessively rare volume of engravings, entitled *Fleurs dessinées d'après nature*, utilizing the technique of stipple engraving that Redouté was later to employ so successfully. Redouté was shortly to combine these two separate aspects of van Spaendonck's art, adding to them his own technical innovations and botanical skills, to achieve the art for which he is now famous. Van Spaendonck also helped Redouté in another, and equally important, way, for he enjoyed an influential position as instructor of flower painting to the *jeunesse dorée* of the Court and aspiring gentry, which led to lucrative commissions. He provided Redouté entrée into these wealthy and privileged circles, adding to the influence already exercised on Redouté's behalf by L'Héritier. The importance of patronage cannot be overstated.

Redouté's first royal appointment was as Designer to the Queen's Chambers, effected by L'Héritier and Desfontaines. The position was not remunerated and its only advantage to Redouté was that it gave him access to the gardens of the Petit-Trianon. In fact Marie-Antoinette detested paintings, and the only time she called upon his services was when, imprisoned in the Temple, she summoned him to paint a night-blooming cactus she was fond of. Louis XVI had been taken prisoner on 10 August 1792, and the Republic was proclaimed six weeks later. The Jardin des Plantes was declared the property of the citizens and renamed the Muséum National d'Histoire Naturelle, and the famous collection of vellums was transferred from the King's Library to the musuem.

In the midst of the tumult of the Revolution, Redouté laboured on. He continued his paintings for the collection of vellums, his position having been made official by the Convention. He was also at work on hundreds of engraved plates for botanical books and journals. Before looking at Redouté's apogee during the Consulate and First Empire, under the aegis of Joséphine, mention should be made of a few of the splendid publications for which he provided illustrations. A chronological recitation would be difficult, for it must be remembered that Redouté was working simultaneously on several projects, and that these costly books were issued in instalments, sometimes taking up to twenty years to complete. L'Héritier brought out his *Geraniologia*, with forty-four engravings of geraniums, of which thirty-three are from drawings by Redouté. Desfontaines's *Flora Atlantica*, with ninety-two of its plates by Redouté, illustrates the plants found during Desfontaines's exploration of Algeria and Tunisia. Ventenat's *Description des plantes nouvelles et peu connues, cultivées dans le jardin de J.M. Cels* and its sequel, *Choix des plantes*, together contain ninety-one plates by our artist. Michaux's *Histoire des chênes de l'Amérique*, illustrating the new species of oak discovered in the eastern region of the United States, has thirty-two plates by Redouté out of a total of thirty-six, while de Candolle's *Astragalogia* contains fifty plates, all by Redouté. The most impressive in size, but alas not in quality, is the *Nouveau Duhamel*, a massive seven-volume work begun in 1800 and dedicated to Joséphine, illustrating all the trees and shrubs in outdoor cultivation in France. By the time of its completion in 1819 it contained 498 colour-printed plates, of which 463 are by Redouté. These are only some of the most prominent works illustrated by him during this time.

Another work which deserves more careful consideration is de Candolle's *Plantarum Succulentarum Historia*, or *Plantes Grasses* as it is known by its French title. It was begun in 1798, and is notable for being the first work illustrated entirely by Redouté and the first to utilize his method of colour printing from stipple-engraved plates. It is also the only book for which Redouté engraved all of the plates himself. It owed its inception again to L'Héritier, whose role in Redouté's career was now sadly diminished; the Revolution had left him penniless and out of favour. L'Héritier had early on conceived of a great work on cacti and other succulent plants, and Redouté had been amassing paintings of them whenever opportunity offered. The importance of illustration for the study of succulent plants is that, unlike most other plants, few succulents can be dried and preserved in a herbarium in recognizable form; therefore illustrations become one of the prime tools of identification. Although L'Héritier could not now sponsor this project, Desfontaines found a young Swiss botanist, Augustin Pyramus de Candolle, to provide the text and an enterprising publisher, Garnéry, to undertake the work. For the first time Redouté had complete control over the preparation of the plates, and from then on he exercised full authority, commissioning text, supervising the engravings, even obtaining subscriptions and effecting sales. Previously most of his drawings had been engraved surrounded by a line frame; this was a tradition established in the collection of vellums, and one insisted on by most botanists. In the *Plantes Grasses*, he dropped the frame, leaving the illustrations free and unencumbered on the page, and he never returned to the convention of the frame. One hundred copies of the folio edition were struck (a quarto edition was also published), and Redouté probably executed the hand finishing on all these copies himself. He had found a technique, inspired by Bartolozzi and van Spaendonck and, however unperfected in the light of his later refinements, set a standard of botanical illustration which led to his great works, those on Malmaison, *Les Liliacées*, and *Les Roses*.

Before considering the above-mentioned works and Redouté's subsequent career, we should recapitulate the history of botanical illustration leading up to the technical breakthrough which Redouté achieved with the *Plantes Grasses*. With the invention of printing in the fifteenth century begins the history of printed book illustration: woodcut ornaments, initial letters, and full-page woodcuts supplant the skills of the manuscript illuminator and illustrator. In the earliest printed books these skills were often combined, the printed text being adorned with hand-painted initials and rubrication, but quickly the mechanical production of these devices took over. The first botanical books were based on illustrated manuscripts, their woodcuts faithfully copying the ornamental inaccuracies of their mediaeval sources. But the dawning of the age of science saw a rapid advance on these crude prototypes: plants were drawn from nature, not merely copied from copies of copies, engraved on wood, and reproduced in book form. Colour was desired for aesthetic and botanical reasons, so the woodcuts were occasionally painted over with gouache or watercolour, sometimes by the publishers (Fuchs' herbal was certainly issued in this way), sometimes by amateurs. With the development of metal engraving, finer detail could be achieved. These illustrations too were often coloured in the same manner: the Dutch florilegia of the seventeenth century, de Pass, de Bry, Sweert, and others, demonstrate the art attainable. By the eighteenth century a pattern is established: engraved prints, usually in folio format, often coloured by hand in watercolour. Some experiments in colour printing were ventured, such as Weinmann's *Phytanthoza Iconographia* (1737–45) and Martyn's *Historia Plantarum Rariorum* (1728–36), utilizing a primitive mezzotint. Usually not more than two, or three at most, colours would be printed, the remainder being added by hand. The effects obtained were crude and not equal to the great hand-coloured engraved books.

While in London Redouté had met the artist Bartolozzi, who had perfected a technique of engraving, utilizing dots and not lines, a style that greatly infuriated William Blake who preferred clean, incisive lines. Bartolozzi found that he was able to apply different colours by hand to the plates and then print from them.

Redouté was impressed and upon his return to Paris found van Spaendonck also working on stipple engraving and experimenting with colour printing. Redouté quickly began his own experimentation; some of the plates in the *Stirpes Novae* are partially colour-printed, although utilizing the traditional line engraving. It was not until the publication of the *Plantes Grasses* that Redouté produced an entire book illustrated solely by colour-printed stipple-engraved plates. Having demonstrated what could be achieved, he was determined to perfect this technique.

Stipple involves the careful and tedious engraving of minute dots, their density varied to effect tone and shading. Line engraving is also used to indicate veining, contours, and other highlights. The engraved plate is then in effect 'painted' with the colours required, the ink adhering to each of the dots. The inking, or colouring, of such a plate is far more difficult than the hand colouring of an ordinary engraving already printed on paper. After printing, each plate would then be thoroughly cleaned and the process repeated for another copy. It should be emphasized that, unlike our modern techniques of colour printing, what Redouté was attempting to achieve was not a more efficient and less expensive form of illustration; his colour-printed plates resulted from probably the most complex reproduction methods possible. Rather, he was aiming to create illustrations with the qualities of luminosity, sheen, and dimensionality which were so strongly in evidence in the paintings on vellum by van Spaendonck and himself. For this purpose, finely engraved stipple was ideal, and the minute dots of colour, against the white ground of paper, created a *pointilliste* effect which a solid wash of watercolour, having uniform tone and hue, could not equal. Even then the printed plate would not be finished, for, however perfect, it would still require details and nuances to be added in watercolour by hand. No wonder that, for *Les Liliacées* alone, Redouté had a team of eighteen engravers working under the direction of the skilled Tassaert, among whom was Langlois, who had already been producing the splendid colour-printed bird plates for the works of Audebert and Levaillant.

Much technical mastery was also involved in preparing inks of the right qualities, so that they would not run or bleed into adjacent colours, and in obtaining the proper paper as well. We can imagine Redouté orchestrating this impressive atelier, finding the best craftsmen, training them to his standards, relentless in his supervision of every detail of the production. He allowed himself no margin for error, he synthesized techniques which separately would be considerable enough, but which together produced the finest series of botanical illustrations ever known. To do this he required patronage on a vast scale.

The rewards of Redouté's association with the Court of Louis XVI were stinted and meagre, but now, with Napoléon in the ascendant he was about to enter the most creative and productive period in his life, a period that was to last over twenty years. Napoléon's wife and empress to be, the lovely Creole Joséphine, bought the estate of Malmaison at Reuil in 1798, and determined to make its gardens a botanical monument, filled with the rare and exotic plants that were being brought back from all parts of the globe. Joséphine appointed Ventenat as official botanist, Mirbel as director and agent, and Redouté as artist, to capture with his pen and brush the living treasures cultivated there. Redouté became Joséphine's protégé and close personal friend—this was a friendship that would survive her later divorce and eclipse. He was given the princely salary of 18,000 francs a year. Joséphine knew how to spend money lavishly on a scale worthy of Caesar's wife and, while some of her extravagances seem only that, for the most part her patronage was well directed and bore scientific and artistic fruit which would have been stunted without it. At Malmaison a virtual botanical industry was established, with teams of gardeners, botanists, and engravers, coordinated by Mirbel and under the command of Ventenat and Redouté. Plant hunters and explorers were commissioned by Mirbel to bring back exotic species, and daily the gardens saw new additions coming in.

Redouté was not tardy in his efforts. He was already at work on several other publications, some of which have been mentioned; but under the aegis of Joséphine he embarked on the three works which represent the summit of his artistic and botanical achievements. In 1802 the first instalment of *Les Liliacées* appeared, with text by de Candolle. Whereas de Candolle's name had appeared as author of the *Plantes Grasses*, and it is under his name that the work is catalogued, in *Les Liliacées* the title-pages bear Redouté's name alone. In 1803 the first instalment of Ventenat's *Jardin de la Malmaison* was published; this two-volume work, completed in 1805, contains 120 colour plates by Redouté, all illustrating the rare plants at Malmaison, and dedicated to Joséphine. After Ventenat's death in 1808, Aimé Bonpland, von Humboldt's botanist companion in his explorations of South and Central America, took Ventenat's place, and in 1812 appeared the first part of his *Description des plantes rares cultivées à Malmaison et à Navarre*, with fifty-two of its sixty-four colour plates by Redouté. Together these three constitute Redouté's finest work. The art of stipple engraving appears in all its perfection, while the delineation is botanically accurate and at the same time imaginative. There is vigour and variety in each of the 658 plates that together they contain. Scientific accuracy is not sacrificed on the altar of aesthetics, nor vice versa; for these works are monuments of science as well, and to this day retain their documentary relevance to plant science.

With the appearance of *Les Liliacées* and the two great books on Malmaison Redouté had effectively completed his career as a botanical illustrator. His artistic career continued, but from now on his work is to be judged as that of a flower painter, a very different attainment. One work spanning both phases is the famous *Les Roses*, begun at Malmaison in 1817. The technical equal of the *Les Liliacées* in every respect, it was and still is Redouté's most immediately popular work. But in many of the plates an elegant licence is apparent, a quality that came to the fore in Redouté's later works. His artistic skill remained unimpared, but his productions now began to be aimed exclusively at the boudoir and drawing room, and for the botanist they hold little interest. With the fall of Napoléon and the Bourbon restoration, a narrower, more selfish age was ushered in. Redouté continued to enjoy court patronage, but this was not the far-sighted, enlightened patronage of Napoléon and Joséphine. Redouté had to trim his sails to a mean-spirited wind; thus the succession of lavish, decorative and, some would say, rather decadent illustrations and paintings. Conspicuous among this later work in demonstrating his still considerable power is the *Choix des plus belles fleurs*, published between 1827 and 1833, containing 144 colour plates. But here is nothing new, and set beside the innovative majesty of *Les Liliacées* the difference is apparent. Flowers are selected not because they are new and exciting, but for their decorative appeal in combination, and we witness a return to the set floral pieces of the Dutch masters Redouté had so much admired as a youth.

Redouté laboured on, in part to avert bankruptcy which continued to threaten him, his fortunes, but not his popularity, partially in eclipse. He was awarded the Légion d'honneur in 1825, presented by Charles X, and Audubon visited him in Paris and paid him homage. This now grand old man had lived through the reign of Louis XVI, the Revolution, the Consulate, Directory and Empire, the Bourbon restoration, and the July revolution of 1830 which saw the ascendancy of the house of Orléans. He had lived through the most turbulent and dangerous epochs, and was accorded honour and respect by the successive factions and regimes. He had as pupils and patrons five queens and empresses of France as well as Marie-Caroline, the Duchesse de Berry. He enjoyed many friends and admirers, and had no enemies that we know of. And, despite the financial handicaps of his old age, and the change of taste that affected his later work, his skill and perseverance never deserted him.

On 19 June 1840 his daughter brought him a white lily to paint. That afternoon a letter arrived from Rémusat, Minister of the Interior, declining a commission for what was to have been Redouté's most ambitious painting, a large uncompleted canvas entitled the 'Triomphe de Flora'. Redouté was found by his daughter in a state of shock and, although he briefly recovered, he died the next day from a cerebral haemorrhage, aged eighty. He was sitting in front of his easel, the lily which he had been painting beside him.

William Patrick Watson is a specialist in natural history books and a director of B. Quaritch Ltd.

REDOUTE'S *LES LILIACÉES*

William T. Stearn

Plant portraits, such as those of Redouté's *Les Liliacées*, which are aesthetically pleasing, botanically accurate and relevantly detailed, blend both art and scholarship. A flower painter aims primarily to produce an aesthetic effect. A botanical artist has the more exacting task of meeting the needs of a science. No artist, however skilled his or her technique, can illustrate satisfactorily a botanical work without specialized instruction and training in observation of details pertinent to that work; their inclusion is an act of botanical scholarship. It is thus not coincidence but a matter of cause and effect, of co-ordinated effort, indeed of happy collaboration, that the greatest masters of plant portrayal have owed no small part of their pre-eminence to a close working association early in their careers with eminent botanists who drew their attention to the intricacies of floral structure and emphasized the necessity of careful draughtsmanship based upon understanding. The history of botanical illustration teems with examples of such mutually advantageous relationships. They include that of Aubriet with Tournefort, of Ehret with Trew, of S. Parkinson with Solander, of Franz and Ferdinand Bauer, the two greatest of botanical artists, with Jacquin, John Sibthorp and Robert Brown, of Redouté with L'Héritier, Sydenham Edwards with Samuel Curtis, Heyland with A.P. de Candolle, W.H. Fitch with W. Hooker, Riocreux with Brongniart and Decaisne. There are others.

Pierre-Joseph Redouté (1759–1840), although so deservedly celebrated and lastingly popular, was unique neither in his skill nor in his botanical importance. He remains unique, however, in the large number of superbly illustrated works published to his credit. *Les Roses* deals only with a single genus, *Rosa*. *Les Liliacées* provides a gallery of decorative and diverse petaloid monocotyledons belonging to Liliaceae, Iridaceae, Commelinaceae, Amaryllidaceae etc. On account of the many species either first named or first illustrated in its eight volumes, that great work retains its botanical importance as well as its aesthetic appeal.

Redouté did not write the text accompanying his illustrations. Even had he possessed the requisite knowledge and scholarship, he would still have been better employed, as he was, to the benefit of horticulture and botany then and now, in drawing, painting, engraving and plate-production. Three celebrated botanists wrote the text for *Les Liliacées*: Augustin Pyramus de Candolle (1778–1841) for Volumes I–IV (1802–1808), François Delaroche (1782–1814) for Volumes V–VII (1809–1813) and Alire Raffeneau Delile (1778–1850) for Volume VIII (1814–1816). New botanical names published in these volumes should be attributed to their authors and not to Redouté.

The citation of the names of botanists, often in an abbreviated form such as 'L.' for Linnaeus and 'DC.' for de Candolle, after scientific names is not widely understood among the general public or even horticulturists. It is primarily a bibliographical device and conveys much information to the initiated. Thus the citation '*Iris susiana* L.' or '*Iris susiana* Linnaeus' is not intended to give any personal credit but to indicate that this name was published by the Swedish naturalist Carl Linnaeus (1707–1778), probably in his most comprehensive botanical work, *Species Plantarum* (1753), as in fact it was, and thus that the plant concerned was already known in the eighteenth century. The citation of the author can be even more informative. For example, *Osmanthus delavayi* Franchet is the name of a widely grown shrub. Adrien René Franchet (1834–1900) was a French botanist who worked in Paris mostly on herbarium specimens collected in Japan and China, particularly those collected by the French missionary botanists J.P.A. David (1826–1900) and J.M. Delavay (1834–1895) between 1868 and 1895 in the interior of China. Thus to a taxonomic botanist the addition of the authority 'Franch' or 'Franchet' to the name *Osmanthus delavayi* indicates that the plant is probably Chinese and became known for the first time late in the nineteenth century and that the type-specimen is in the herbarium of the Muséum National d'Histoire Naturelle, Paris, all of which is true; Franchet published the name *Osmanthus delavayi* in 1886 and based it on a specimen collected by Delavay in Yunnan, western China.

In addition to providing information, the author's name may also be necessary to avoid ambiguity as when two authors have used the same name for different plants.

The above are simple citations. Botanists also use double citations, e.g. '*Chasmanthe aethiopica* (L.) N.E. Br.' or '*Chasmanthe aethiopica* (Linnaeus) N.E. Brown'. The name within parenthesis, here Linnaeus, is that of the botanist who published the epithet adopted, here *aethiopica*, but who did not publish the specific name as a whole. The parenthesis indicate that he published the epithet in a different context. The basis of the accepted name, technically called the basionym, is in this instance *Antholyza aethiopica* published by Linnaeus in 1759; hence the plant was known in the eighteenth century; the epithet *aethiopica* indicates that it came from southern Africa. The name of the botanist outside the parenthesis, here N.E. Brown, is that of the one who made the combination by transferring the epithet from its original use, here under *Antholyza*, to the generic name adopted, here *Chasmanthe*. This means that the Kew botanist Nicholas Edward Brown (1849–1934) considered that the plant should be removed from the genus *Antholyza* where Linnaeus had placed it and transferred to another genus, i.e. *Chasmanthe*, but that the Linnaean epithet *aethiopica* should be retained for it. The initials of the author are added when there exist two or more authors with the same surname, e.g. J.E. Brown, R. Brown, R.W. Brown and S. Brown, to avoid confusion.

The *International Code of Botanical Nomenclature* (latest edition, 1978), which is periodically revised at international botanical congresses, provides principles, rules and recommendations for the naming of plants. It is a lengthy and complicated legal work intended to promote uniformity and stability, although strict application of the rules to some nomenclatural situations has had, at least for the time being, a contrary effect. The basic principle is that of priority of publication, with Linnaeus's *Species Plantarum* (1753) taken as the starting point, because this work was the first comprehensive one to use the now accepted binomial system of nomenclature consistently. For example, the species for which Linnaeus provided the convenient binomial (two-word name) *Antholyza aethiopica* in 1759 had early been named *Gladiolus aethiopicus flore coccineo* by Cornut in 1635 and *Hyacinthus africanus foliis Colchici, floribus coccineis* by Plukenet in 1691, but these lengthy pre-Linnaean names have no present validity though much historic interest. The principle of priority means that, when there exist two or more different names published since 1753 for the same plant, the correct name is the earliest validly published one in accordance with the accepted classification and the various provisions of the *International Code of Botanical Nomenclature*. Thus, the Boat Lily, Oyster Lily, or Oyster Plant from Central America was named *Tradescantia spathacea* by Swartz in a publication, *Prodromus*, issued in June or July 1788 and *Tradescantia discolor* by L'Héritier in a publication, *Sertum Anglicum*, issued in December 1788 or January 1789. Because the epithet *spathacea* has priority over *discolor* and the species is placed in a genus, *Rhoeo*, distinct from *Tradescantia*, its correct name is *Rhoeo spathacea* and not *Rhoeo discolor*. Similarly the name *Crocus flavus* Weston (1771) has now been adopted on grounds of priority instead of *Crocus luteus* Lamarck (1791), *Crocus aureus* Smith (1806) and *Crocus lageniflorus* Salisbury (1806), all names applied later than this.

Such conflicts of names have often resulted from a person in one country being unaware of a publication issued in another country on the same group of plants. They may also result from differences of opinion on classification. Some plants which now appear to be members of one very variable species, needing only one specific name and that the earliest, may have seemed previously, before intermediate specimens were available, to be an assemblage of several species each needing its individual specific name. The reverse may also be true; then the name formerly used with a wide application is restricted to species represented by the original material and other names are found or coined for the other species.

For such reasons the scientific names given to Redouté's plants by his three highly competent botanical collaborators, de Candolle, Delaroche and Delile, and used on the plates and in the text of *Les Liliacées* are not necessarily those accepted for the same plants today. Of those included in the present volume of a hundred and eight plates, as many as fifty-six names have remained unchanged. Changes in generic allocation have affected twenty-one species; for example, the *Convallaria japonica* of Thunberg is so markedly different from *Convallaria majalis*, the Lily of the Valley, that it has had to be placed in another genus, *Ophiopogon*, and its name has accordingly become *Ophiopogon japonicus*. There have also been changes in specific epithets resulting from the use of the earliest one available in accordance with the *International Code of Botanical Nomenclature*; this has affected twenty-three names here; for example, *Canna latifolia* in place of *Canna gigantea*. In a number of instances the name has totally changed on account of a different generic allocation and the use of an earlier epithet, for example *Hymenocallis rotata* in place of *Pancratium disciforme*, *Alpinia zerumbet* in place of *Globba nutans*. Thanks, however, to the excellence of both text and plates, the identity of plants depicted in *Les Liliacées* is rarely in doubt.

Petaloid monocotyledons resemble succulent plants in that significant features clearly evident in a living state may become obscured or obliterated when specimens are pressed and dried for the herbarium. Good illustrations are thus invaluable records particularly if associated with reliable text. The choice of plants for illustration in *Les Liliacées* was haphazard as was the sequence of plates. Evidently when one came into flower and caught Redouté's attention he painted it and then asked his botanical collaborator to check its identity and write a descriptive text. The diversity of plants thus portrayed reflects the richness of the gardens available to Redouté. They provided material surpassing in quantity those of the other groups to which he devoted his talents. *Les Liliacées* accordingly stands as the highest peak in the mountain range of his achievements as a botanical artist.

A number of plants illustrated by Redouté being then considered species new to science and thus without names were published with new scientific names in *Les Liliacées*. It has subsequently been found that some of these received other names elsewhere. Hence to ascertain the dates of publication of such competing names is necessary for a decision as to which name should be adopted and which rejected as a synonym. About 1806 a low-growing lily with numerous very narrow leaves and scarlet flowers came into cultivation from northern Asia and was immediately recognized as a new species which was illustrated by Redouté, named *Lilium pumilum* by de Candolle in Paris, *Lilium linifolium* by Hornemann in Copenhagen and *Lilium tenuifolium* by Schrank in Munich. Redouté's plate 378 with de Candolle's description has been ascertained to have been published in 1812, which gives the name *Lilium pumilum* priority over the others published respectively in 1813 and 1822. Such considerations led B.B. Woodward to investigate the dates of publication of *Les Liliacées* and to publish his conclusions in the *Journal of Botany, British and Foreign* 43:26 (1905). These are as follows:

Volume I, parts 1–3, pls 1–18, in 1802; part 4, pls 9–24, in 1802 or 1803; parts 5–8, pls 25–48, in 1803; parts 9–10, pls 49–60, in 1804.
Volume II, parts 11–17, pls 61–102, in 1804; parts 18–20, pls 103–120, in 1805.
Volume III, parts 21–23, pls 121–138, in 1805; parts 24–27, pls 139–162, in 1806; parts 28–30, pls 163–180, in 1807.
Volume IV, parts 31–34, pls 181–204, in 1807; parts 35–40, pls 205–240, in 1808.
Volume V, parts 41–46, pls 241–276, in 1809; parts 47–50, pls 277–300, in 1810.
Volume VI, parts 51–58, pls 301–348, in 1811; parts 59–60, pls 349–360, in 1812.
Volume VII, parts 61–67, pls 361–402, in 1812; parts 68–70, pls 403–420, in 1813.
Volume VIII, parts 71–74, pls 421–444, in 1814; parts 75–78, pls 445–468, in 1815; parts 79–80, pls 469–486, in 1816.

Les Liliacées exists in two states. The ordinary folio edition, with plates measuring roughly 20 in. (50 cm) by 13 in. (33 cm) began publication in parts in 1802, as noted above. A large-paper edition, *Edition grand papier*, with plates measuring roughly 24 in (60 cm) by 18 in. (45 cm), began publication in 1807 but was soon issued concurrently with the ordinary folio edition. Concerning this large-paper edition, Redouté himself states that 'chaque Liliacée sera retouchée par moi au pinceau, après l'impression'. A signed pencil note by him in the Lindley Library copy says that there were only eighteen such copies, 'dix-huite seulement'.

The 108 plates selected for reproduction out of the total 486 include all the true lilies, i.e. members of the genus *Lilium*, illustrated by Redouté as well as representatives of such other genera as *Asphodeline*, *Erythronium*, *Fritillaria*, *Tulipa*, belonging to the family Liliaceae, and likewise all representatives of such other genera as *Chasmanthe*, *Crocus*, *Gladiolus*, *Sparaxis*, belonging to the family Iridaceae. They also include decorative representations of other families such as the Amaryllidaceae, Butomaceae, Commelinaceae, Musaceae, Zingiberaceae. These plates are typical of *Les Liliacées* as a whole. By displaying beautifully and instructively the great diversity of habit and flower within the petaloid monocotyledons, Redouté's 'Liliacées', they happily exemplify the union of art and scholarship marking the work of the great masters of botanical illustration.

An English naturalist, Martin Lister (*c.* 1638–1712), wrote in 1682, 'Naturall History is much injured, through the little incouragement which is given to the Artist, whose noble performances can never enough be rewarded, being not only necessary, but the very beauty and life of this kind of learning.' Fortunately such 'noble performances' had become more appreciated a century later when Redouté began to draw flowers in the Jardin du Roi of Paris. The encouragement and patronage he subsequently received there have one of their finest monuments in *Les Liliacées*; here indeed is 'the very beauty and life of this kind of learning'.

Professor Stearn, botanist and bibliographer, is President of the Linnean Society, London and of the Garden History Society.

Some Sources of Further Information

Blunt, W.J.W., 1950. *The Art of botanical Illustration*, London (Collins)

Delcourt, R. & Lawalrée, A., 1949. 'Pierre-Joseph Redouté, botaniste illustrateur (1759–1840)', *Lejeunia* 13:5–20

Lawalrée, A, & Buchheim, G., 1972. *P.J. Redouté*, Pittsburgh (Hunt Institute for Botanical Documentation)

Lawrence, G.H.M. (*Ed.*), 1963. *A Catalogue of Redouteana exhibited at the Hunt Botanical Library*, with contributions by F.A. Stafleu, I. MacPhail, J.V. Brindle and A. Lawalrée, Pittsburgh (Hunt Botanical Library)

Léger, C., 1945. *Redouté et son Temps*, Paris

Sitwell, S. & others, 1954. *Album de Redouté*, with contributions by S. Sitwell, R. Madol and W.T. Stearn, London (Collins)

Stearn, W.T., 1973. 'The principles of botanical nomenclature, their basis and history'. In P.S. Green (*Ed.*) *Plants wild and cultivated*, p.86–100. Hampton, Middlesex (Classey, for the Botanical Society of the British Isles)

Woodward, B.B., 1905. 'Redouté's works', *Journal of Botany* 43:26–30.

SOME BIOGRAPHICAL NOTES

Michel Adanson (1727–1806). A French botanist who worked on a natural system for the classification and naming of plants, which is described in *Familles des plantes* (1763). For four years, from 1749, he worked in Senegal, where he collected many plants and in 1757 he published a survey of the flora of that country.

Adam Afzelius (1750–1837). This Swedish botanist founded the Linnaean Institute in Uppsala in 1802 and later became professor of botany at the university. Earlier he had visited London and through the recommendation of Sir Joseph Banks had been sent on a collecting expedition to investigate the plant resources of Sierra Leone. His records for this trip were published in *Genera Plantarum Guineensis* (1804).

William Aiton (1731–1793). The early history of Kew Gardens contains an impressive array of names, one of the most famous being that of William Aiton, who was appointed Head Gardener in 1759. He had been trained at Chelsea Garden by the great gardener Philip Miller. His work at Kew probably reached its peak with the publication of his *Hortus Kewensis*, containing descriptions of 5,600 exotic plants that had been introduced into Kew or neighbouring gardens.

Sir Joseph Banks (1743–1820). A great and wealthy explorer and naturalist famed for his patronage and promotion of science. He began his impressive career at an early age, visiting Newfoundland and Labrador in 1766. He sailed in *The Endeavour* with Captain Cook, returning with a vast number of specimens and later, at the invitation of George III, he became honorary director of Kew Gardens. It was largely through Banks and the then Head Gardener, William Aiton, that Kew began to acquire its status as one of the world's leading botanical gardens. Banks was instrumental in sending out many expeditions, including that of Captain Bligh in *The Bounty* to introduce Bread Fruit to the West Indies.

Augier Ghislain de Busbecq (1522–1592). A Flemish diplomat who was sent by the Emperor Ferdinand of Austria as ambassador to the court of Suleiman the Magnificent. During this period he travelled in Turkey and gathered a fine collection of manuscripts, including a fifth-century copy of Dioscorides' *Materia Medica*, which bears the name of the princess Juliana Anicia, daughter of the Emperor Flavius Anicius Olybrius. Busbecq also sent plants from Turkey to Vienna and has the distinction of introducing the garden tulip into western Europe.

Carolus Clusius (1526–1609). A student not only of plants but of law, philosophy and theology, Clusius was also an accomplished Latin scholar. Plants, however, dominated his life and in 1573 he was appointed director of the imperial garden in Vienna by Emperor Maximilian II. He travelled widely in Europe and his journey in the Iberian Peninsula provided the material for his famous *Rariorum Stirpium per Hispanias Historia* (1576). His later years were spent teaching in Leiden and it is said that his introductions and cultivation of tulips during this time led to the great Dutch bulb industry.

Pedanios Dioscorides (born c. 50 AD). A Greek medical man, Dioscorides was born in Cilicia, southern Turkey and served in Nero's army. His work *De Materia Medica* (c. 70 or 80 AD) was extremely popular in the Middle Ages. It gives details of about 600 medicinally useful plants grouped, basically, into the aromatic, the nutritional and the medicinal, although some are arranged according to their natural affinities. During the Renaissance Dioscorides' work was so highly regarded that only those drugs described by him were considered to be genuine.

John Gerard (1545–1612). An English herbalist famed for his popular book, *The Herbal or General History of Plants* (1597). More than 1,000 species are included in it, with details of their habitats, flowering times, uses and connections with folklore. It is now useful as a source of information on what plants were cultivated in English gardens at that time. Unfortunately, much of Gerard's work was not original but was taken from that of the Flemish botanist R. Dodoens and from T. Tabernaemontanus' illustrations in *Eicones Plantarum.*

Antoine Laurent de Jussieu (1748–1836). Member of an extraordinary family of French naturalists, five of whom made their mark on botanical history. Antoine Laurent developed the foundations of a natural system of plant classification, published in his *Genera Plantarum.*

Carl Linnaeus (1707–1778). The work of this Swedish naturalist formed the present system of classification of living creatures. At the age of twenty-three he was lecturing to students at the botanic garden in Uppsala, and at this time began publishing papers and studying the classification of the plant world. He travelled widely in Europe and in 1741 was appointed professor of botany at the university of Uppsala, where in the next twenty years he built up the botanic garden to be the finest in Europe. From there he trained and sent out his students to many parts of the world. His great collection of books and dried specimens were bought by Mr J.E. Smith of Norwich for 900 guineas and these are now in the care of the Linnean Society of London.

Matthias de Lobel (1538–1616). A Flemish botanist who studied at Montpellier and later worked in England, during which time his *Stirpium Adversaris Nova* was published, written in collaboration with his colleague Petro Pena. This was followed by several other works which were highly esteemed in the botanical world. He described many new plants and adopted a system of classification which represented a definite advance on those of earlier botanists.

John Parkinson (1567–1650). A herbalist, and apothecary to King James I, Parkinson is noted for his publication *Paradisi in sole Paradisus Terrestris* (1629). This famous work containing descriptions of nearly 1,000 plants led to his being bestowed with the title 'Botanicus Regius Primarius' by Charles I. The wood blocks for the figures in this work were specially cut in England. Later he published *Theatrum Botanicum* (1640) in which 3,800 plants were mentioned.

C.S. Rafinesque (1783–1840). This brilliant naturalist and writer was born in Turkey but his name is linked more closely with America, for in 1818 he was appointed professor of botany, natural history and modern languages at a university in Lexington. In 1825 he created a botanic garden there. His main work involved the study of a natural system of classifying plants, which differed markedly from that of Linnaeus.

R.A. Salisbury (1761–1829). The botanist Salisbury was the son of Richard Markham, a cloth merchant of Leeds. He took the name of Salisbury when in 1785 a wealthy acquaintance of his grandmother, a Miss Anna Salisbury, gave him £10,000 to follow his studies in botany and gardening. In 1806 he published *Paradisus Londinensis* with drawings by his friend and colleague William Hooker after whom he named the genus *Hookera.* Undoubtedly Salisbury indulged in much plagiarism but at the same time he was a careful and accomplished botanist.

B.M.

The Plates

When you have only two pennies left in the world,
buy a loaf of bread with one and a lily with the other.

Chinese saying

LILIACEAE

Lilium bulbiferum Linnaeus

ORANGE LILY

Among the European lilies, this is easily distinguished by its erect orange or reddish flowers. As long ago as 1597 John Gerard observed that 'there have been from the beginning great controversies about the name of this red Lillie . . .', but the acceptance by H.B.D. Woodcock and W.T. Stearn (1950) of the Orange Lily as one species with several varieties is probably as near to the truth as one can get and is certainly practical from the horticultural viewpoint. *Lilium bulbiferum* variety *bulbiferum*, the typical form, grows to a height of about 1 to 1.5 metres and has flower segments of a deep orange-red on the inside, with a central patch of orange and a deeper red exterior. The scattered alternate leaves sometimes have bulbils in their axils, but this seems to be a variable feature and of little importance in distinguishing between the varieties. *Lilium bulbiferum* variety *croceum* (*Lilium aurantiacum* of Richard Weston) is the well-known Orange Lily, or Feuerlilie, which is so commonly seen in gardens and has become the emblem of the Orangemen of Northern Ireland because it flowers at the time of the Orange Festival on 12 July. Its flowers are lighter orange, slightly darker at the tips of the segments and orange externally, thus lacking any of the reddish tinge of variety *bulbiferum*. Variety *chaixii* of G. Maw is a dwarf form from the Maritime Alps, with only a few saffron-coloured flowers which have reddish-orange tips to the segments. It is recorded by H.J. Elwes, however, that plants of this variety grow larger when brought into cultivation and it may be stunted only because of its rocky natural habitat. Variety *giganteum* is, as its name suggests, a robust many-flowered form up to 2 metres in height, which was collected near Naples. The distribution of *Lilium bulbiferum*, including all its variants, is a wide one—central and southern Europe, from the Pyrenees to Poland and Hungary.

1. Pistil 2. Capsule 3. Bulb

Lilium Bulbiferum

LILIACEAE

Lilium candidum Linnaeus

MADONNA LILY

Probably the best-known lily of all, *Lilium candidum* has a classical history second to none in the plant world. It is figured, for example, on Assyrian bas reliefs of about 700 BC, but long before then—some 3,500 years ago—it was used as an ornament on a bronze dagger of the Middle Minoan period. Its popularity since has continued unabated and it has long been associated with the Christian Church. In the eleventh century the Navarre Knights of the Order of Our Lady of the Lily fought in Spain with *Lilium candidum* embroidered on their breasts, and fifteenth-century artists such as Lippi and Botticelli often used the white lily as a symbol of purity to enhance their religious paintings. The name Madonna Lily is of comparatively recent origin and may date from as late as the nineteenth century. It was cultivated not only for its unquestionable beauty but also for its healing properties and was used, for example, in an ointment for softening corns or tumours and for reducing areas of inflammation.

Apart from its fascinating history, *Lilium candidum* is something of a botanical curiosity. The whitish scaly bulbs produce a set of basal leaves in autumn which remain green all through the winter, quite unlike the majority of species, which remain firmly dormant during this inclement period of the year. The flower stem, which can eventually attain 2 metres, appears in spring and reaches its zenith at flowering time in June or July.

The cultivation of the Madonna Lily is by no means simple. In some gardens it will thrive with little by way of assistance and it is often a feature of old cottage gardens; in others no amount of tender care will yield success. Sun and a lime-rich soil are two factors that do seem beneficial to its growth. Because of its long history of cultivation, the wild habitat of this lily is not known with certainty, but in the eastern Mediterranean region it can be found in apparently natural situations growing in rocky limestone areas of scrub, where its roots are shaded but the flower stems are in sun. In many regions it is undoubtedly naturalized in areas around historical sites.

Leaves of lower stem

Lilium Candidum

LILIACEAE

Lilium chalcedonicum Linnaeus

Of the European species of lily there are few with red Turk's-cap-shaped flowers, and among these it is a very simple matter to distinguish *Lilium chalcedonicum*, the only species to have, as it were, two different sets of leaves, all of which are strikingly silver-margined. The lower ones are rather normal in their appearance, being lanceolate and spreading, while the upper are quite different—much smaller and pressing against the stem. This clearly separates the Chalcedonian Lily from the other red-flowering species such as *Lilium pomponium*, *Lilium carniolicum* and the red form of *Lilium pyrenaicum*.

The name *chalcedonicum* is taken from the former district of Chalcedon opposite Istanbul and the species was introduced to Europe from Constantinople in the time of Queen Elizabeth I. It was in fact known to John Parkinson in 1629 as 'the Red Martagon of Constantinople', Martagon being an ancient name for lilies of this type.

Although the plant was brought at an early date into European gardens from Istanbul, as were many other beautiful bulbous species of course, it appears to be rather more typically a native of Greece and may have been known in Turkey as an introduced plant.

It is now a rather rare species, recorded growing in northern Greece, Euboea and the Peloponnese in rocky places in woodland clearings. Needless to say, such a striking plant did not go unnoticed by the ancient Greeks and it is clearly illustrated in Minoan artwork.

Redouté illustrated *Lilium chalcedonicum* twice, because it was first labelled as *Lilium pomponium* (page 31) and when the mistake was noted, a correction was made in the text for the second illustration. He considered that the later specimen was remarkable for its solitary flower, and in fact called it 'variety A' of *Lilium chalcedonicum*, but this feature is of little importance in terms of classification. The species varies very little in general appearance but can have from one to ten flowers per stem depending upon the vigour of the specimen.

Lilium Chalcedonicum

LILIACEAE

Lilium chalcedonicum Linnaeus
figured as *Lilium pomponium*

1. Bulb 2. Stamens and pistil shown in relation to one perianth segment

Lilium Pomponium

LILIACEAE

Lilium martagon Linnaeus

TURK'S-CAP LILY, MARTAGON LILY

This common European lily with its sharply recurved perianth segments has given its name to a whole group of lilies bearing similarly shaped flowers. They are not confined to Europe and Asia Minor as is *Lilium martagon* itself, but are distributed right across Asia and North America, although some authorities are inclined to separate the American species because of their rather different bulb structure.

The curious name *martagon* is thought either to be derived from a similar Turkish word for a turban of a particular shape, or to have some connection with the planet Mars. In mediaeval times plants of medicinal value were often considered to be under the influence of a star, and G.C. Wittstein suggests that the name comes from *Marti genus*, child of Mars. Whichever is the true explanation, the common name 'Turk's-cap Lily' has become almost as firmly established as 'Martagon Lily' and is usefully descriptive.

Although in its most common colour forms *Lilium martagon* might perhaps be described as dowdy by comparison with many other lilies, it compensates by being extremely easy to grow. It will naturalize in grass or between shrubs and so is ideally suited to a semi-wooded garden.

Over its whole distribution, from Portugal in the West, eastwards to Russia and northern Turkey, it varies somewhat in appearance, and at one time in the nineteenth century up to twenty different forms were known, but most of these are not now in cultivation. The most attractive variants to be seen in gardens now are variety *album* with the white, unspotted flowers and variety *cattaniae* from Dalmatia—sometimes known as variety *dalmaticum*—which has striking dark burgundy-wine-coloured flowers, also unspotted.

Capsule

Lilium Martagon

LILIACEAE

Lilium canadense Linnaeus
figured as *Lilium penduliflorum*

CANADA LILY

This common and widespread lily of eastern North America was almost certainly the first species from the continent to be introduced into cultivation in Europe. The French are thought to have brought it from there in the 1620s and John Parkinson described it in 1629 as 'the spotted Martagon or Lily of Canada'. Today it is known locally as the Canada Lily or Wild Yellow Lily.

Lilium canadense is a most distinguished plant, with whorls of leaves and up to twenty large pendulous bell-like flowers whose tepals curve gracefully outwards but never recurve as do those of the true Martagon Lily and its kin. Over its enormous range of distribution it has some variation in colour, and there are orange and red variants as well as the yellow. Most of these forms have flowers which are spotted inside with brown but variety *immaculatum* is of a clear unmarked yellow. Some of the other variations have also been given names: for example, variety *flavo-rubrum*, a deep orange; variety *editorum*, red with narrow tepals; and variety *coccineum*, the deepest brick-red form.

In the wild, *Lilium canadense* is a mid- to late-summer-flowering plant, occurring on the edges of woodlands and in damp meadows from Quebec southwards to West Virginia. In cultivation it therefore requires a humus-rich soil in slight shade and is one of the most attractive of wild lily species for growing in acid conditions such as in rhododendron beds and peat gardens.

Bulb

Lilium Penduliflorum

LILIACEAE

Lilium philadelphicum Linnaeus

GLADE LILY, WOOD LILY

The history of this brightly coloured, rather delicate lily is well documented in the very fine work *Lilies of the World* by Hubert B. Drysdale Woodcock and William T. Stearn (1950). It appears that it was first introduced to cultivation in 1675 in Paris by Monsieur Richer, who sent it from Acadie, a French colony consisting of what is now Nova Scotia and part of New Brunswick. Not until nearly a hundred years later was it named *Lilium philadelphicum* by Linnaeus, who, although giving the distribution as Canada, apparently used the name of this American state because John Bartram of Philadelphia had sent bulbs to the famous gardener Philip Miller in the early part of the eighteenth century.

A beautiful slender little plant, it is usually not more than 1 metre in height and sometimes as low as 15 centimetres. It is distributed from the eastern states of Quebec and Ontario southwards to North Carolina. Rather surprisingly, however, this species which is so widespread and successful in the wild is extraordinarily difficult to grow in gardens. The natural habitat is in dryish woods or grassland, not, it seems, in particularly specialized conditions, but rarely is *Lilium philadelphicum* seen in cultivation.

Over this wide area many colour variants exist, with flowers ranging from yellow to deep red and either unspotted or intensely black-spotted. Orange, however, seems to be the most common colour form encountered. The local common names are delightful and numerous, a selection being Glade Lily, Huckleberry Lily, Wild Orange Lily and Wood Lily.

The difficulty of cultivating *Lilium philadelphicum* in England is rather succinctly stated by that well-known authority, P.M. Synge, who describes it as 'a perpetual challenge to keen growers'.

1. Perianth segment, showing inner surface
2. A stamen 3. Pistil

Lilium Philadelphicum

LILIACEAE

Lilium pumilum de Candolle

This is one of the most dainty of all the lily species, certainly among the Martagon or Turk's-cap types. The slender wiry stems have fairly numerous, very narrow leaves, and this feature, together with its dwarf stature separates *Lilium pumilum* from all other similar species. The stems rarely reach more than 45 centimetres and in June carry up to twenty of the sealing-wax-red flowers. The form illustrated by Redouté seems a rather weak specimen; it is usually a little more robust in appearance than this.

In the wild, *Lilium pumilum* is a native of northeastern Asia—China, Mongolia, eastern Siberia, Korea and Tibet. Indeed, clumps of the bulbs are grown by the Tibetans in the mud-built roofs of their houses, where they flower freely. It is not a difficult lily to cultivate in Europe and can be quickly raised from seed to flowering size. The bulbs, however, often die after flowering, so a continuous supply of young bulbs should be raised to keep a succession going.

It appears that *Lilium pumilum* was introduced into cultivation in the gardens of Russia in the early nineteenth century and arrived soon after in western European horticulture, for it was described by A.P. de Candolle in 1812 in *Les Liliacées*. The Redouté illustration shows a curious solid-looking bulb, which is perhaps slightly misleading although it is true that the species has a bulb with rather few white scales compared to the very scaly bulbs of many liliums.

In addition to the usual bright scarlet-flowered form, there are yellow-flowered variants, Golden Gleam being one of the most well known.

Lilium Pumilum

LILIACEAE

Lilium pyrenaicum Gouan

PYRENEAN LILY

One of the easiest lilies for garden purposes, *Lilium pyrenaicum* is a native of the Pyrenees and northern Spain, where it is a plant of light woodland and mountain meadows. So tolerant is it that in Britain it has become established in hedgerows in a few places. It is ideally suited to a wild garden situation or for naturalizing in grass and seems to thrive on almost any soil, from acid sands to chalky downlands.

This is a true Turk's-cap lily, with the strongly reflexed perianth segments of the Martagon group and, although the individual blooms are not large, a vigorous clump of specimens with up to twelve flowers on each plant can be really striking.

Among the European lilies, *Lilium pyrenaicum* in its typical form is distinguished by its leaves being hairless on the underside and in having dark-spotted yellow flowers which are much larger than those of the Balkan *Lilium albanicum*. There is a reddish-orange form of *Lilium pyrenaicum*, known as variety *rubrum* and this has at times been confused with *Lilium pomponium*, which has in turn been confused with *Lilium chalcedonicum*. From *Lilium pomponium* the Pyrenean Lily can be distinguished by its wide leaves, up to 2 centimetres broad, and by the more orange-red shade of the flowers. The former has leaves not more than 1 centimetre wide and flowers of a brilliant sealing-wax red. *Lilium chalcedonicum*, illustrated on pages 29 and 31, differs from both of these in having the upper stem leaves much smaller than the lower ones and held against the stem. The only other species to be considered with regard to the relationships of *Lilium pyrenaicum* is the southeast European *Lilium carniolicum*; although orange-flowered too, this last has leaves that are hairy on the underside.

1. Pistil, with a stamen and a perianth segment
2. Leaves of the lower stem

Lilium Pyrenaicum

LILIACEAE

Lilium superbum Linnaeus

SWAMP LILY

This late-summer-flowering American Swamp Lily is well named 'superb', for it is not only large-flowered and showy but can also carry as many as forty of the graceful pendulous blooms on robust stems up to 3 metres tall. It is easily grown in, as its common name suggests, damp places, for in the wild it is a plant of moist meadows and marshes on acid soils. Thus it associates well with rhododendrons and moisture-loving primulas in semi-woodland situations and can be ranked as a thoroughly desirable garden plant. The bulbs are stoloniferous, so that there is a tendency for the plants to wander somewhat from their original planting position.

Its range of distribution in the United States is from Massachusetts southwards to Florida and it thus occurs in a more easterly region than the similar Californian *Lilium pardalinum*. Unlike the latter species, which is orange right into the centre of the flower, *Lilium superbum* has flowers with a green star-shaped centre formed by the furrowed nectaries. The other quite distinctive feature of *Lilium superbum* is the shape of the flower bud, which is very angular with three flat surfaces. In its truly Turk's-cap-shaped flowers with their conspicuously reflexed segments, *Lilium superbum* differs from the related *Lilium canadense* (page 35), the flowers of which curve outwards at their tips but do not reflex.

Lilium superbum has probably been known in European gardens for about two hundred and fifty years, for it is on record that it was introduced in the early eighteenth century.

Several colour variants of the Swamp Lily are known, including ones with clear yellow flowers, rich red flowers and unspotted forms with reddish tips to the segments.

1. Capsule 2. Pistil and stamens

Lilium Superbum

Lilium lancifolium Thunberg
figured as *Lilium tigrinum*

TIGER LILY

It is very unfortunate that the well-known name *Lilium tigrinum* for this species has to be replaced by the earlier name *Lilium lancifolium* based by C.P. Thunberg on an imperfect specimen with bulbils on the stem but no flowers. The names *Lilium tigrinum* and Tiger Lily, as also Tigridia, refer to the jaguar, still known in South America as a 'tiger' and not to the striped Asiatic tiger.

Lilium lancifolium is, next to *Lilium candidum*, one of the best known of all lilies and has been cultivated in the Far East for at least a thousand years, partly as a food. It seems that it did not reach Europe until 1804, when bulbs arrived at Kew from Canton. China, Korea and Japan are the regions of the Orient in which it is probably truly wild, although it is difficult to be sure of the native provenance of a plant that is frequently cultivated and increases at a prodigious rate by bulbils in the axils of the leaves.

Really vigorous plants can reach 2 metres in height and in late summer carry up to twenty-five large spotted blooms, their segments gracefully reflexed. This is a very distinctive lily with its stout, dark purplish stems covered with cobweb-like hairs, and blackish-purple axillary bulbils. Unfortunately, of all the lilies, it is one of the most susceptible to viruses and it rarely makes impressive patches before succumbing to a crippling disease.

There are various forms of *Lilium lancifolium* that have been given varietal status: variety *flaviflorum* for example, with its yellow flowers spotted with dark purple. Variety *splendens*, considered one of the best variants, has many larger, strongly spotted flowers of a more reddish orange, while variety *flore-pleno* must rate as a curiosity only, having double flowers with about two to three dozen narrow segments and no stamens. Variety *fortunei* was introduced to the West in 1861 from Japan and is a fine robust plant with up to fifty large orange flowers and a very hairy stem.

A stamen

Lilium Tigrinum

LILIACEAE

Agapanthus africanus (Linnaeus) Hoffmannsegg figured as *Agapanthus umbellatus*

The genus *Agapanthus* is well known in European gardens for its large umbels of blue flowers in summer. Although entirely confined to southern Africa in the wild, several of the ten species have proved to be excellent and hardy garden plants in the northern hemisphere and there are now many highly ornamental hybrids.

The species are distributed in both the southwest Cape region of South Africa, which receives winter rainfall and the eastern Cape, which has rain in summer. Those *Agapanthus* from the former region are evergreen while those from the summer-rainfall areas die down in winter to a resting crown.

Agapanthus africanus belongs to the evergreen group and this, together with the fact that it occurs between sea level and an altitude of only 1,000 metres, means that it is one of the less hardy species in the gardens of northern Europe. It is therefore best treated as a cold greenhouse or conservatory plant in areas that are subjected to heavy frosts. Another extremely satisfactory way of growing it is in ornamental containers which can be placed on a patio after the danger of frosts has passed, for it can provide a long display of huge bright blue umbels of flowers from mid-summer to autumn.

It is recorded that this species was cultivated at Hampton Court gardens in 1692, but it was certainly already in cultivation nearly two decades earlier than this in Holland.

The *Agapanthus* were first regarded as tuberous-rooted hyacinths and only in 1788 did Charles Louis L'Héritier establish the genus, the name of which is derived from the Greek *agape*, love, and *anthos*, flower.

Flower opened out to show pistil and stamens

Agapanthus Umbellatus

Allium moly Linnaeus

GOLDEN GARLIC

The genus *Allium*, which includes onions, garlic, chives and leeks, is made up of a very great number of species—one could probably count around two hundred and fifty in Europe and Asia Minor alone without considering the rest of northern Asia and North America. Among so many species it is a relief to find one that is instantly recognizable. Neither a skilled botanical eye nor the aid of a lengthy key are needed to identify *Allium moly*, which differs from all the others in having yellow flowers in a rather shuttlecock-shaped, or hemispherical, umbel and lanceolate grey leaves, usually about 2 to 3.5 centimetres wide. A truly bulbous species, it has a somewhat rounded bulb, and this too makes it distinctive, since most of the other yellow-flowered 'onions', and there are comparatively few in the genus, have elongated or bottle-shaped bulbs. *Allium scorzonerifolium* is the nearest relative to *Allium moly* but can easily be distinguished by its linear leaves less than 1 centimetre wide and its angled and crested ovary.

The Golden Garlic has been a popular garden plant for several centuries, and quite justifiably so, for it is brightly coloured, neat in appearance and easy to cultivate. It flowers in early summer and can be used for rock gardens and borders, or planted beneath shrubs. The bulbs increase naturally by division.

No doubt it was the golden flowers of *Allium moly* rather than the cost of the bulbs that made this flower a symbol of prosperity in gardens of the distant past.

In the wild it is a native of southwest France (eastern Pyrenees) and eastern Spain, where it inhabits shady rocky places.

1. Perianth segment with stamen 2. Capsule
3. Bulb

Allium Moly

LILIACEAE

Allium nutans Linnaeus

This robust species of onion is an attractive one, with its tufts of grey-green leaves and dense spherical umbels of rosy pink or purplish flowers. It belongs to the section Rhizirideum of this enormous genus, a group characterized by the narrow, somewhat bottle-shaped bulbs which tend to be clustered on a short rhizome and therefore have a clump-forming habit. In *Allium nutans* the stout flower stems are 30 to 60 centimetres tall and have two-winged ribs in the upper part. Although the fully expanded umbel is upright, in the bud stage it is bent over and drooping; hence the epithet *nutans*. The leaves are up to 1.5 centimetres broad—unusually wide for members of this group—and flat on the underside, a feature that distinguishes the species from the similar *Allium angulosum*, which has a sharp keel on the lower surface of the leaves. The better-known European *Allium senescens* is very closely allied to *Allium nutans*, but tends to be rather smaller in stature. It is in the stamens, however, that the major differences can be found. In the former species the filaments are simple, with no teeth, and only slightly longer than the perianth segments. In *Allium nutans*, on the other hand, each filament is often furnished with two teeth and is at least half as long again as the perianth segments.

Allium nutans is a native of Siberia and central Asiatic Russia, where it inhabits steppes, meadows and stony slopes.

1. Flower with stamens 2. Pistil

Allium Nutans

Allium carinatum – Linnaeus – subspecies *pulchellum* Bonnier & Layens

figured as *Allium paniculatum*

Fortunately it is possible to divide the very large genus *Allium* into several sections according to various features of the underground parts, leaves, fruits and flower structure. *Allium carinatum* subspecies *pulchellum* belongs to the section Codonoprasum of H.G.L. Reichenbach in having an egg-shaped bulb, leaves sheathing the stem, two spathes, which are unequal and exceed the umbel of the flowers by long tail-like appendages, and flowers which are usually cup-shaped rather than flat or starry.

As can be seen from the illustration, the stamens protrude considerably from the reddish-purple flowers and these two features make the species immediately distinguishable from the related *Allium pallens* and *Allium paniculatum* in which the flowers are pale brown, yellowish or pale lilac and have the stamens more or less concealed within the flower. It is also related to the well-known *Allium flavum* and to *Allium stamineum*, but the former has yellow flowers and the latter has rather pale, purplish or yellowish-brown flowers and a more or less globose ovary, whereas *Allium carinatum* subspecies *pulchellum* has an ovary which is much longer than it is wide.

This attractive plant is much cultivated in gardens, sometimes in an albino form, under the name '*Allium pulchellum*'. W.T. Stearn, however, regards it as a subspecies of the Keeled Garlic, *Allium carinatum*, which normally carries bulbils in the umbel, sometimes to the exclusion of flowers. Subspecies *pulchellum* is a native of grassy places in southeast France and eastwards to Romania and Greece, whereas subspecies *carinatum* is more widespread in Europe and is naturalized in several northern European countries such as Britain.

Allium Paniculatum

LILIACEAE

Asparagus maritimus (Linnaeus) Miller
figured as *Asparagus amarus*

WILD ASPARAGUS

Unlike the curious *Asparagus asparagoides* (page 83), the species illustrated here is a much more recognizable member of the genus and looks rather like the cultivated vegetable, asparagus, which is a form of *Asparagus officinalis.*

The species in this worldwide genus number some three hundred and are classified largely according to the structure of their cladodes, the leaf-like organs that replace the true leaves, which are reduced to tiny scales or spurs at the base of the cladodes. In some cases the cladodes are very spiny, but on the whole they are rather pine-needle-like, certainly in the European species.

Asparagus maritimus is a native of the central and eastern Mediterranean region, where it inhabits sandy areas usually near the sea. Its herbaceous stems grow to a height of 1 metre and are erect and self-supporting. These stems have ridges of tiny papillae running along their length, and this is one of the characteristics whereby the *Asparagus maritimus* can be distinguished from the common *Asparagus officinalis* the stems of which are smooth. Another distinguishing feature is that the flowers of *Asparagus maritimus* are carried on rather short stalks, usually less than 1 centimetre in length, whereas the pedicels of *Asparagus officinalis* are normally between 1 and 2 centimetres long.

Asparagus Amarus

LILIACEAE

Asphodeline lutea (Linnaeus) Reichenbach
figured as *Asphodelus luteus*

YELLOW ASPHODEL

The genus *Asphodeline* was at one time included with *Asphodelus*, but in 1830 H.G.L. Reichenbach split the latter and formed a new genus for the species illustrated here and some related ones. The two genera look rather unalike and need not be confused, for in *Asphodelus* the leaves are all produced in a basal tuft from the apex of the root system, whereas in *Asphodeline* the leaves are carried on the flowering stem. Although *Asphodeline lutea* and a few other species in the genus *Asphodeline* have yellow flowers whilst the *Asphodelus* species are without exception white- or pinkish-flowered, the two genera cannot safely be distinguished by the colour of the flowers because there are white-flowered species of *Asphodeline* too. As far as *Asphodeline lutea* (the subject of this plate) is concerned, however, it could only possibly be confused with its close relative *Asphodeline liburnica*. To distinguish between these two, one need only observe the position of the leaves on the stem, for the former has leaves all the way up as far as the flowers, while the latter has leaves clustered only on the basal half of the stem.

The Yellow Asphodel, or King's Spear, is a native of the central and eastern Mediterranean region, and is often very common on dryish rocky hillsides in the Balkans, Asia Minor and North Africa. The yellow flowers, produced in summer in rather dense racemes, are at their best in the evenings, tending to close up during the heat of the day. For this reason the asphodels do not perhaps make very spectacular garden plants. The rootstocks consist of short rhizomes from which radiate long fleshy yellow roots not unlike those of the cultivated asparagus.

1. Perianth segment with stamen 2. Pistil

Asphodelus Luteus

LILIACEAE

Colchicum autumnale Linnaeus

MEADOW SAFFRON

Although the plant figured by Redouté was given the varietal name of *latifolium* in the original text, there seems to be no valid reason for doing so, since this common colchicum is quite a variable plant and the leaves illustrated are not excessively large. It is, however, rather unusual for *Colchicum autumnale* to have as many as six leaves per corm as shown in the plate. A more remarkable feature of the particular plant chosen by Redouté is the small size of the flowers, but since they were apparently 'produites par une bulbe plantée presque à la surface de la terre' (produced by a bulb planted very near the surface of the ground), it is possible that they are not as fully developed as they might be in this species.

Colchicum autumnale, the Meadow Saffron, is very widespread in Europe as far east as the Ukraine, but does not apparently reach Asia Minor. It flowers in late summer or autumn and can often be seen growing by the million in the meadows of central Europe. The leaves are produced long after the flowers are over, usually not until the spring, and the rather denuded appearance of the flowers has given rise to the local name of Naked Boys. The name Autumn Crocus for this colchicum is very misleading, because it is of course not in any way related to *Crocus*. Indeed, any link with the true Saffron Crocus through the common names is a dangerous one, since the colchicum is extremely poisonous whereas saffron is used for culinary purposes and *Crocus* corms are frequently eaten raw by shepherds and villagers in the Middle East.

Colchicum autumnale contains colchicine, a substance long known as a powerful poison but which has medicinal value too if utilized correctly. It is also much used in plant breeding experiments to induce polyploidy—chromosome changes—in other plants.

1. Whole fruit   2. Fruit split open
3. Transverse section of fruit   4–6. Seeds

Colchicum Autumnale

Colchicum agrippinum Baker
figured as *Colchicum variegatum*

Colchicum agrippinum is somewhat of a mystery plant and we can only guess that it is a garden hybrid of considerable antiquity. This supposition is based on the fact that it is absent in the wild and also that it is sterile, a condition often indicative of hybrid origin. Moreover, it bears marked similarities to two other species, *Colchicum variegatum* and *Colchicum autumnale*, both of which were certainly in cultivation as long ago as the early seventeenth century. It seems likely that a hybrid between the two would produce a plant similar to that figured here.

Whilst stating on the one hand that *Colchicum variegatum* originated in the Island of Chios, Redouté also remarks on the fact that the subject actually used for the plate came from the garden of Monsieur Cels. Both *Colchicum variegatum* and *Colchicum agrippinum* were in cultivation at this time and clearly there was some confusion between the two.

Exactly how the name *Colchicum agrippinum* arose is not known, but it was certainly in use long before J.G. Baker validated it in 1879. The most probable explanation is that the plant was being grown, or perhaps even originated, in Cologne. This city was known to the Romans as Colonia Agrippinensis and it may well have been an association of the plant with the city that gave rise to the curious epithet.

Colchicum variegatum, a native of the Greek Islands and southwestern Turkey, has flattish flowers with spreading perianth segments which are very strongly and clearly chequered. The leaves are undulate, grey-green and often prostrate. *Colchicum autumnale*, on the other hand, has non-chequered funnel-shaped flowers and erect green leaves. Between the two with regard to these features, *Colchicum agrippinum* has widely funnel-shaped flowers with rather poorly defined chequering and erect green leaves with slightly undulate margins. The leaves of all three plants are produced in winter and spring, long after the flowers are finished.

Colchicum agrippinum, though perhaps not as beautiful as *Colchicum variegatum*, is a much more satisfactory garden plant, increasing rapidly by corm division and flowering freely.

Colchicum Variegatum

Ophiopogon japonicus (Thunberg) Ker-Gawler figured as *Convallaria japonica*

DWARF LILY-TURF

This little, exceedingly tough, grass-like plant belonging to the Liliaceae has been widely used for ground cover and even as a kind of turf, although not the kind that can be mown. This latter usage has led L.H. Bailey to suggest the name 'Lily-Turf' for the group of species belonging to the genera *Liriope* and *Ophiopogon*. To summarize their usefulness one can do no better than quote Bailey when he writes, 'They make extensive ground-covers, withstanding heat, drought and shade; they are durable pot plants for porches and living rooms. They thrive in warm and tropical countries and at least one of them stands in the open as far north as Boston.' (*Gentes Herbarum*, Vol. 2, 1929, p. 3.)

Ophiopogon japonicum, the Dwarf Lily-Turf, is one of the smallest species in the whole group and makes very dense patches of growth by means of subterranean stolons. Although the small drooping flowers are not showy, being very pale lilac or whitish and not very freely produced, the fruits are an attractive blue, nearly 1 centimetre in diameter and persist for several months on the plant. The species has a long history in cultivation, for John Bellenden Ker-Gawler records in the *Botanical Magazine* of 1808 that the plant was introduced into Kew Gardens in 1784. It was, however, known to botanists much earlier than this, and Engelbert Kaempfer described and illustrated it in his *Amoenitatum Exoticarum* of 1712. To Bailey it was known as *Mondo japonicum*, but although the generic name *Mondo* of Michel Adanson (1763) is earlier than Ker-Gawler's, the later name *Ophiopogon* has been conserved under the International Rules and is therefore the valid one. In the original text for Redouté's illustration, de Candolle followed C.P. Thunberg in placing it in the genus *Convallaria*, which is nowadays restricted to one species, the Lily of the Valley.

1. Flower seen from above (actual size) 2. Flower seen from below (actual size) 3. Flower with perianth removed to show stamens 4. Same, enlarged 5–6. Stamens (enlarged) 7. Pistil with base of perianth 8. Fruit 9. Seed

Convallaria Japonica

Convallaria majalis Linnaeus

LILY OF THE VALLEY

In spite of several specific names in the genus *Convallaria*, it is most likely that there is in fact only one exceedingly widespread and not very variable species spread right across temperate Europe, Asia and North America, although it is possibly naturalized in the latter continent.

With its broad leaves, slender rhizomatous underground parts and raceme of small globose scented flowers, the charming Lily of the Valley bears little resemblance to any other member of the lily family. The Solomon's Seal, *Polygonatum*, is probably one of the nearest relatives, but its flowers are more tubular or bell-shaped than those of *Convallaria*, and the stems bear several scattered leaves.

In former times *Convallaria majalis* was used for medicinal purposes: a bitter-tasting extract from the roots proved effective as a purgative, while the dried and powdered flowers were used as snuff and an infusion of the flowers was recommended for headaches. However, it is as a delightful garden plant that the Lily of the Valley has been most appreciated, providing as it does an effective low ground cover for woodland or shaded conditions. Moreover, the deliciously scented flowers are excellent for cutting and nowadays very early flowers are commercially produced by freezing the rhizomes and then forcing them into growth in gentle heat.

There are several variants of *Convallaria majalis*, notably a pink-flowered form called Rosea and one with richly golden-striped leaves, as well as a robust garden variety known as Fortin's Giant, and Prolificans which produces more than one flower on each pedicel so that the racemes have many more flowers.

Convallaria Majalis

LILIACEAE

Cordyline fruticosa (Linnaeus) A. Chevalier figured as *Dracoena terminalis*

The wild form of *Cordyline fruticosa* is not a particularly showy plant, with its green leaves and pale pinkish flowers. However, the species has been subjected to much selection and a host of differently coloured leaf forms make it highly prized as an ornamental foliage plant. In temperate countries it can be grown under glass or as a house plant and some of the forms are among the most brilliant of all foliage pot plants. The leaves vary in width from lincar to broadly elliptic and may be green, yellow, purple or pink, or any combination of these colours, usually in stripes or zones. The flowers on most forms of this plant have little decorative appeal, but in some tropical countries there are varieties which have more eye-catching mauve, reddish or white flowers followed by showy red berries.

In regions of tropical Asia where *Cordyline terminalis* is a native, it is often used as a hedge plant around houses, to ward off evil spirits and bring good luck. The leaves too have special significance in folklore: worn around the neck they are thought to impart divine power to priests and make warriors invulnerable. In the countless uses to which this species has been put, hardly any part of the plant has been ignored. In Hawaii it is recorded that the leaves have been used to make hula skirts, sandals, raincoats, and as thatching material, wrappers for food, cattle fodder and toboggans for children to slide down grassy slopes. The roots, which are fleshy and bulky, can provide large quantities of rather sweet food or, fermented and distilled, can be made into a type of brandy.

The origin of the variants with highly coloured leaves is a little obscure. Certainly several of those known in Europe were introduced by James Gould Veitch in the late nineteenth century, so it would appear that at that time the South Sea Islanders had already made selections.

Cordyline fruticosa itself is a shrub, or even a small tree, reaching about 4 metres in height at maturity, with a bare stem showing many leaf scars. As a cultivated pot plant it is normally not allowed to develop to this extent, although the plants are of no commercial interest when very young since the leaf colours do not develop properly for a few years.

Dracoena Terminalis

Erythronium dens-canis Linnaeus

DOG'S TOOTH VIOLET

The Dog's Tooth Violets, allegedly thus named because of the similarity in shape between their bulbs and canine teeth, belong to a smallish genus with an interesting distribution. There are about twenty-one species, almost every one exquisitely beautiful. The species *Erythronium dens-canis* covers the whole area from western Europe eastwards through the Balkans, the Caucasus and Iran to Siberia and Japan. Its appearance varies slightly from place to place and because some botanists have considered it possible to distinguish several of these variants, other names are sometimes attached to them, bearing geographical epithets such as *Erythronium sibiricum*, *Erythronium caucasicum* and *Erythronium japonicum*.

However, it is in the New World that conditions have apparently favoured the further development of the genus *Erythronium*. Apart from *Erythronium dens-canis* which is so widespread in the Old World, all the remaining species are North American. About fourteen belong to the western states and some of these are very limited in their distribution and rather rare.

The European Dog's Tooth Violet is a delightful spring-flowering plant for gardens, especially when naturalized in grass or beneath shrubs in a semi-wild setting. In its native haunts it is a plant of light woodland, often on limestone formations, and flowers soon after the snow has receded in spring.

1. Perianth segment and single stamen 2. Pistil

Erythronium dens Canis

LILIACEAE

Eucomis comosa (Houttyn) Voss
figured as *Eucomis punctata* L'Héritier
PINEAPPLE FLOWER

The name *Eucomis* is derived from two Greek words *eu* and *kome* meaning, literally, 'good hair' or, by extension, 'beautiful head', which obviously refers to the tuft of leaves surmounting the flower spike in these plants. The common name Pineapple Flower, or Wildepynappel in Afrikaans, is very apt, for this cluster of 'leaves' do indeed bear some resemblance to the leaves that overtop the fruit of the pineapple. In *Eucomis* they are in fact not true leaves but leaf-like sterile floral bracts. The reasons for this evolutionary step are not at all obvious but, whatever its cause, this feature makes *Eucomis* instantly distinguishable from all the other genera of the lily family.

There are about twelve to fifteen species of *Eucomis*, all African and most of them confined to the eastern Cape regions in Natal, Lesotho, the Transvaal and the Orange Free State. The species illustrated here grows in marshy places, along streams and sometimes in woods, where it flowers from December to February, that is, in mid-summer. In northern hemisphere gardens its flowering time is usually June, July and August, again the summer months. Although perhaps not strikingly beautiful, the *Eucomis* certainly have considerable interest value and are attractive hardy bulbs for a sunny border. Some of the hybrids with pink flowers and purplish leaves are quite ornamental and the group could undoubtedly be much improved as garden plants by selection and hybridization.

Eucomis punctata of Charles Louis L'Héritier is evidently the same as Martin Houttuyn's *Asphodelus comosus* which, being an older name, takes priority, making the plant figured here *Eucomis comosa*. However, the genus is in need of careful revision before the names of all the species can be decided with certainty.

Eucomis Punctata

LILIACEAE

Fritillaria imperialis Linnaeus

CROWN IMPERIAL

The stately Crown Imperial is by far the most spectacular and graceful of all the fritillaries and it is not surprisingly one of the earliest plants on record to be cultivated. Its introduction to Europe dates back at least to Matthew de L'Obel in the sixteenth century, but no doubt it was used as a garden plant in Asia well before this. Like so many of the excellent plants that have beautified our western gardens *Fritillaria imperialis* probably first reached Vienna by way of Istanbul, that venerable trading centre of the horticultural world.

The pendent bells contain a drop of glistening nectar at the base of each perianth segment and legend has it that the plant hangs its head in shame and sheds tears of repentance because it once refused to bow its head as Jesus passed on the way to Calvary. The cluster of brick-red bells is unusual in the genus and there are only a few species which carry their flowers in this way.

The Crown Imperial is a native of southeast Turkey and eastwards through Iran and Afghanistan to the Himalaya range, inhabiting sunny rocky mountainsides. This gives an indication of its requirements for successful cultivation in gardens, and a well-drained sunny position is a good start. There are several colour variants of *Fritillaria imperialis*, including yellow forms and various shades of orange or brick-red. The related *Fritillaria eduardii* from Russian central Asia has rather wider flaring bells of orange, while the lovely though less robust *Fritillaria chitralensis* from the western Himalayas has a few clear yellow bells. *Fritillaria raddeana* from Turkmenistan and northeast Iran has smaller greenish-yellow flowers. Together these constitute within the genus a distinct group known as the section Petilium; all have their flowers in an umbel overtopped by a cluster of narrow leaf-like bracts.

Fritillaria Imperialis

LILIACEAE

Fritillaria meleagris Linnaeus

SNAKESHEAD LILY

Apart from its stately and quite different-looking relative the Crown Imperial, *Fritillaria meleagris* is probably the best known of the fritillaries, with its strongly chequered reddish-purple flowers.

The generic name is thought to have been derived from 'fritillus', the sixteenth-century name for a chessboard and in *Fritillaria meleagris* one can clearly see why, for the mottling is squarish like a tessellated floor. The present English name of Snakeshead Lily may have something to do with the way in which the buds are carried, rather like a snake rearing up to strike, but more probably this name refers to the likeness of the chequering to the scales on a snake's head. The old name in England, Ginny-hen Flower, also calls attention to the flower pattern, likening it in this case to the markings on the guinea fowl, *Numida meleagris*. Both these vernacular names are infinitely more attractive than Leper Lily, which refers to the similarity in shape between the flower and a leper's bell.

Fritillaria meleagris is a very widespread European species from France and Switzerland eastwards to Russia and north to Britain. Inhabiting moist meadows, it is easy to grow and has certainly been a popular garden plant for at least four hundred years. It has even been suggested that the Romans brought the plant into England for culinary purposes and that it is now distributed around their former settlements.

The most common form of *Fritillaria meleagris* is reddish-purple, but pinkish and white forms occur quite frequently in wild populations.

Pistil and stamens

Fritillaria Meleagris

Fritillaria persica Linnaeus

PERSIAN FRITILLARY

The Persian Fritillary is rather more widespread in its distribution than its name suggests and occurs not only over a very wide area of Iran but also westwards to Iraq, Syria, Lebanon and Turkey. It varies greatly in the colour of the flowers, from a rather dirty straw-yellow to brownish and deep plum purple, the latter form being by far the most attractive from the gardener's point of view. In recent years a really superb form, of robust constitution and producing many blackish-purple flowers, has been distributed commercially under the name of Adiyaman, a town in southern Turkey where it was collected. Similarly coloured forms do however occur in other parts of its range and it is not possible to generalize and assume that Turkish forms are more richly coloured than those from elsewhere.

Taxonomically, *Fritillaria persica* has no obvious relatives in the genus and it can be regarded as a very distinct and rather isolated species.

It appears to have been one of the first species in this attractive genus to be brought from Persia to Constantinople (now Istanbul) and from there on into other parts of Europe. We can be certain that it was known in western gardens, as *Lilium persicum*, in the late sixteenth century and it may well have been grown for several decades before this. It is perhaps surprising that it has enjoyed such popularity for it is scarcely one of the most pleasing of the fritillaries, with the habit of growing tall and leafy with small flowers. Indeed, in 1636 John Gerard was moved to remark, 'We have in our English fields many scores of flowers far exceeding it'!

1. Perianth segment, showing nectary
2. Stamens and pistil 3. Pistil

Fritillaria Persica

LILIACEAE

Hemerocallis fulva (Linnaeus) Linnaeus

DAY LILY

This is the best known of all the Day-Lilies and has been in cultivation for several centuries. The curious fact is that the exact origins of the plant are unknown and while it appears to occur in wild situations in such diverse localities as France and China and Japan, it seems highly probable that it has been distributed as vegetative propagations by man. It is a very robust but self-sterile plant, which spreads into large clumps by means of stolons and is readily divided and increased in this way. One can imagine how easily it could have been dispersed throughout the gardens of Europe, China and Japan.

Since most of the twenty or so species are Asiatic, it is quite likely that *Hemerocallis fulva* has its origins in that part of the world and that the very widely distributed form is a self-sterile clone which will not produce seeds unless crossed with another species or with a different clone of the same species. Various forms of it are known, including a double-flowered one, and some variants have longer perianth tubes than the one illustrated here by Redouté. The colour is normally reddish-brown or orange but sometimes there are bands of reddish-purple on the segments, one such variation having been given the name of variety *maculata*.

At the present time, there are numerous colourful garden Day-Lilies, but whether they are selections of individual species or hybrids between two or more is not on record. What can be said is that they are first-class summer-flowering herbaceous border plants, whatever the complications of their taxonomy!

Hemerocallis Fulva

LILIACEAE

Lachenalia bulbifera (Cyrillo) Ascherson & Graebner figured as *Lachenalia pendula*

The large South African genus *Lachenalia* is named after Werner de Lachenal, 1736–1800, who was professor of botany in Basel. The large number of species vary greatly in appearance, from those with long tubular pendulous flowers such as the species illustrated here to ones with short erect flowers in very dense spikes. Nearly all colours are represented among the species, some having bi- or tri-coloured flowers, while others break into a range of almost indescribable metallic bluish shades. The fleshy leaves may be spotted or not and vary from species to species not only in shape and colour but also in texture, some having curious pustules all over the surface.

Lachenalia bulbilifera was amongst the first of Cape plants to be introduced to Kew Gardens, collected in 1774 by Francis Masson. It was beautifully figured in the *Botanical Magazine* of 1801 and during the nineteenth century, together with other species, became popular as a conservatory plant. Lachenalias are easily grown and, being natives of the southwest Cape region, flower in winter or early spring, which makes them exceedingly useful as pot plants.

The species figured here is one of the best known in cultivation and increases rapidly by bulb division. In its native South Africa it usually grows at the edge of coastal woods at low altitudes, often in the sandy areas near the sea, where it flowers from June to August.

1. The three outer perianth segments 2. The three inner perianth segments, pistil and stamens 3. Pistil

Lachenalia Pendula

Asparagus asparagoides (Linnaeus) Druce
figured as *Medeola asparagoides*

The seemingly ridiculous name *Asparagus asparagoides* can be explained by the fact that the plant's first valid name under the International Rules of Nomenclature was *Medeola asparagoides* of Linnaeus. It was later considered by W.F. Wight that the species did in fact belong in the genus *Asparagus*, but the specific epithet had to stay, giving the somewhat confusing combination of words meaning 'Asparagus-like Asparagus'!

This unusual South African member of a large worldwide genus, possibly consisting of some three hundred species, is not altogether typical of the plants we think of under the name asparagus: firstly it is a climber and secondly it has broad 'leaves'. These features have undoubtedly been the reason for the considerable confusion the plant has caused among botanists. Over the years it has been placed in at least six different genera, including *Ruscus* and *Dracaena*. Today, however, it is generally accepted that *Medeola asparagoides*, the name under which it was figured by Redouté, does in fact belong to the genus *Asparagus* which, together with a few related genera, is normally placed in the lily family, and by some authorities in the family Asparagaceae.

Asparagus 'leaves' are considered to be distinct from true leaves and are named cladodes; the true leaves are much reduced and bract-like. In many species the cladodes are slender and rather like pine needles, but in the plant illustrated here they are broadly ovate to lanceolate and much more leaf-like.

Asparagus asparagoides is common in the southern and eastern regions of tropical Africa where, from a cluster of swollen tubers, the plant annually sends up stems to a height of 3 metres. It normally twines or scrambles through other vegetation, although occasionally it is self-supporting.

1. Flower, enlarged 2. Perianth segment and single stamen 3. Pistil

Medeola Asparagoides

Gloriosa superba Linnaeus
figured as *Methonica superba*
GLORIOSA LILY

Although the Gloriosa Lilies are spread throughout tropical and southern Africa and India and vary a great deal in their appearance from place to place, it is very difficult to find any logical way in which to classify the variations. The extremely diverse opinions about the status of these variants range from the view that there is only one species, *Gloriosa superba*, with many forms, to the view at the other extreme according to which over thirty different species are recognized.

Although many of the different forms are tall and climb by means of leaf-tip tendrils, there are also miniature ones, such as *Gloriosa minor* from northern Kenya, the stems of which are self-supporting, carry very narrowly linear leaves and reach only 20 to 30 centimetres in height.

Such beautiful plants have understandably received the attention of gardeners for many centuries and since they are very easily propagated by division of the tubers, there has been no difficulty at all in perpetuating the various forms. Apart from differences in stature and the fact that some are climbers and some non-climbers, *Gloriosa* varies enormously in flower shape and colour. The perianth segments may be crisped or undulate on the margins, or straight, and they may be reflexed or curled right back like those of *Lilium martagon*. The colour ranges from yellowish-green to pure yellow, from tomato-red to deep red or purple and the flowers are often bi-coloured. During the life of an individual flower the colour can change from yellow to red, so that the whole inflorescence may be multi-coloured. It is easy to see how such names as 'glorious' and 'superb' came to be heaped upon these amazing plants, but it was partly as a reaction against this repetition of superlatives that such botanists as A.L. Jussieu preferred to use the generic name *Methonica*!

1. Capsule 2. Capsule cut to show seeds 3. Seed

Methonica Superba

Ornithogalum nutans Linnaeus

DROOPING STAR OF BETHLEHEM

The large genus *Ornithogalum* consists of about one hundred and fifty species spread throughout southern and tropical Africa, Europe and the Mediterranean region and eastwards to Afghanistan and adjacent Russian central Asia. There is little in common between the African species and those of the northern hemisphere and there is scarcely any cause for confusion between the species of the two regions. However, the representatives of Europe and Asia do present considerable problems of identification. Nearly all the species from these areas have white starry flowers with green-backed perianth segments, although there are a few from Spain with completely white flowers and one or two with a wholly greenish coloration.

Ornithogalum nutans, the Drooping Star of Bethlehem, is one of the more distinct European species and together with *Ornithogalum boucheanum* forms such a distinct group within the genus that J.G. Baker was moved to create for these two alone the subgenus *Myogalum*. These two spring-flowering species are distinguishable from the rest in having a rather one-sided raceme of widely bell-shaped flowers and fruits that are quite fleshy and pendulous. The filaments of the inner three stamens have a conspicuous pointed appendage on either side of the anther. The structure of the filaments is a rather important character in the genus *Ornithogalum*; although it does not have a marked effect on the overall appearance of the flower it is very diagnostic and must not be overlooked. Fortunately with *Ornithogalum nutans* and its relative, the silvery-white widely flaring flowers with a greyish-green exterior make recognition very simple without the need for closer examination. To distinguish between the two species a detailed investigation is however necessary. In both species the mid-vein of the filament is furnished with a raised crest; in *Ornithogalum boucheanum* this is toothed, whereas in *Ornithogalum nutans* it has no teeth and is smooth.

1. Flower 2. Stamens 3. Pistil 4. Capsule

Ornithogalum Nutans

LILIACEAE

Paris quadrifolia Linnaeus

HERB PARIS

The Herb Paris, Herb True Love or One-Berry belongs to a small group of related genera sometimes placed in a separate family known as Trilliaceae, *Trillium* being the main genus in the group. Unlike this predominantly American genus, which has some representatives in Asia, *Paris*, with about twenty species, is European and Asiatic. The flower parts and leaves in *Trillium* are strictly in threes, whereas *Paris quadrifolia* has four leaves in a whorl and the perianth similarly has four sepals and petals. The latter are very narrowly linear and unobtrusive, the most obvious parts of the flower being the wider sepals and prominent stamens. Other species of *Paris* have their leaves and flower parts in whorls of more than four. The green, scented flowers attract and are pollinated by flies.

Paris quadrifolia is a very widespread species in central and northern Europe, eastwards to Siberia. It is a plant of damp woodlands, where its slender rhizomes wander amongst other vegetation sending up at irregular intervals the leafy flowering stems.

A decoction of the leaves, known as Herba Paridis, has been used as a headache cure and for neuralgia, gout and rheumatism, while the dried powdered rhizomes are reputed to be an efficient emetic. Also it is said that the juice of the blackish, berry-like fruits is beneficial for inflammation of the eyes and that it has properties similar to those of opium.

1. Flower 2. Stamen 3. Fruit 4. Capsule cut to show seeds 5. Seed

Paris Quadrifolia

LILIACEAE

Anthericum liliago Linnaeus
figured as *Phalangium liliago*

ST BERNARD'S LILY

The very large genus *Anthericum*, consisting of perhaps over two hundred species, is mainly African in distribution, although a few species do grow in Europe, America and Asia. Very little is known of the majority of them as garden plants, but the hardy European ones, such as the species figured here, have been cultivated since at least the sixteenth century, for John Gerard mentions them in 1596.

Anthericum liliago is a summer-flowering perennial which dies down in winter, so that it makes an ideal herbaceous plant. It will reach about 40 to 50 centimetres in height when in flower with racemes of starry white flowers not unlike those of *Ornithogalum*, or Star of Bethlehem. However there is little else to connect the two, for *Anthericum* has a cluster of long fleshy white roots while *Ornithogalum* is a bulbous plant. The nearest relative to *Anthericum* is *Paradisea liliastrum*, the St Bruno's Lily, but any confusion between them is more likely to arise from the similarity of the specific and vernacular names than from the appearance of the two plants. *Paradisea* also has racemes of white flowers, but they are bell- or funnel-shaped and much larger than those of *Anthericum*, which are flattish and starry in appearance.

Anthericum liliago is a common plant in Europe from Sweden southwards and extends into Turkey and North Africa. It usually inhabits dryish grassy places but may also be found on rocky slopes and in light woodland. Of this large genus, only two other species occur in Europe: *Anthericum ramosum*, which is also widespread, and *Anthericum baeticum*, which is confined to damp meadows in the mountains of southern Spain. Compared with the St Bernard's Lily, both of these have small flowers, which are carried in a much-branched inflorescence in the former and on very short stalks in the latter. *Anthericum liliago* usually has an unbranched inflorescence or, very occasionally, one or two short branches, and the flowers are carried on stalks 1 centimetre or more long.

Redouté's illustration clearly represents *Anthericum liliago*, but the bracts subtending each flower are unusually long for the species; normally they are much shorter than the stalks, or pedicels.

Phalangium Liliago

LILIACEAE

Polygonatum multiflorum (Linnaeus) Allioni

SOLOMON'S SEAL

The Solomon's Seal is said to bear this curious name because of the seal-like marks which are left on the rhizomes after the leafy stems have died away in the autumn. This seems more convincing than the alternative explanation according to which the rhizome, if cut transversely, bears hieroglyphic-like markings supposedly representing the seal of Solomon, who was knowledgeable in the uses of plants.

A decoction of the roots in wine used to be given to those with broken bones, and preparations from polygonatums were reputed to have the property of healing broken skin. This history has led to the further suggestion that the name 'seal' was connected with the sealing of wounds. Other uses recorded for this serviceable plant include the treatment of lung complaints and haemorrhoids, while the distilled juice has been utilized in the cosmetic industry as a skin-cleaning agent. Moreover, when cooked, the young shoots are said to taste rather like asparagus.

Polygonatum is a large genus, widespread in the temperate regions of the northern hemisphere. There are about five species native in Europe, one of which is *Polygonatum multiflorum*. This plant is distinguished from the closely related *Polygonatum odoratum* in having up to six flowers in a cluster in the axils of each of the lower leaves. *Polygonatum odoratum* only ever has one or two flowers, these, as its name suggests, being sweetly scented whereas those of *Polygonatum multiflorum* are unscented.

1. Fruits 2. Fruit cut to show seeds

Polygonatum Multiflorum

Sansevieria hyacinthoides (Linnaeus) Druce
figured as *Sanseviera guineensis*

By illustrating this species as *Sansevieria guineensis*, Redouté was following the misconception dating from a century or more earlier, that the plant was introduced into cultivation in Europe from Guinea.

Most of the sixty or so species of *Sansevieria* belong to Africa, though there are a few outlying ones in Arabia, India and Madagascar. They constitute a very distinct genus, easily recognized by the very leathery fibrous leaves, often banded with different shades of green, and the tough rhizomes which enable the plants not only to spread into large colonies, but also to survive long periods of adverse climatic conditions. Because of this ability to survive, certain species have come to be known as almost indestructible house plants.

Sansevieria hyacinthoides is one of the few South African species; it grows in wooded areas on hills up to about 1,500 metres, mainly in the coastal southeastern Cape region from around Port Elizabeth eastwards to the East London area.

In his review of the genus in the *Kew Bulletin* of 1915, N.E. Brown writes that a factory was established in Grahamstown (presumably in the nineteenth century) to manufacture fibre from the leaves of this species, but it did not prove successful. However, there are species that are still used as fibre plants on a local scale by some African tribes. Another use for sansevierias is in folk medicine, a decoction of the rhizomes being taken as a treatment for intestinal parasites.

This species was introduced into cultivation in Holland in the seventeenth century and was illustrated without flowers in Commelin's *Horti medici Amstelaedamensis Plantarum Descriptiones et Icones* (1701) and with flowers in his *Praeludia Botanica* (1703). The latter illustration is now accepted as the type of the names *Aloe hyacinthoides* and *Aloe hyacinthoides* variety *guineensis* of Linnaeus published in 1753, and hence of the names *Sansevieria hyacinthoides* and *Sansevieria guineensis* based upon these. The name *Sansevieria thyrsiflora* of C.P. Thunberg, published in 1794, is a synonym. The name *Sansevieria hyacinthoides* has also been used for a species of Sri Lanka illustrated by Commelin in 1701 without flowers, Linnaeus's *Aloe hyacinthoides* variety *zeylanica*, now named *Sansevieria zeylanica*. The confused nomenclature of these species has been elucidated by D.O. Wijnands in *Taxon*, Vol. 22 (1973).

1. Flower opened out to show stamens 2. Pistil

Sanseviera Guineensis

LILIACEAE

Hyacinthoides non-scripta (Linnaeus) Chouard ex Rothmaler figured as *Scilla non-scripta*

ENGLISH BLUEBELL

This very common and well-loved plant of western European woodlands has been the subject of much discussion regarding its true position in relationship to *Scilla*. By some botanists it has been considered a true *Scilla*, by others a *Hyacinthus*, and it has also resided in *Endymion* and *Hyacinthoides*. As if this confusion were not enough it has also received several different specific names, so that quite a number of combinations are possible. The latest opinion, let us hope the last and nearest to the truth, is that our beloved Bluebell belongs to a small genus of only three or four species, which is closely related to *Scilla*. It may be distinguished from *Scilla* by the following characters: the small long-pointed bracts at the base of each flower stalk are paired, not solitary as in *Scilla* and the bulb, instead of consisting of several concentric or imbricated scales, is solid and is completely replaced by a new one each year. In common with those of *Scilla*, the perianth segments are not joined to each other to form a tube as they are in certain related genera with blue flowers such as *Hyacinthus*, *Chionodoxa* and *Puschkinia*.

The only species that bears any resemblance to *Hyacinthoides non-scripta* is the Spanish Bluebell which is in nature confined to western Spain and Portugal but is widely cultivated in gardens. It is a stouter plant with larger more widely flaring bell-shaped flowers which are carried all around the stem, not in a one-sided raceme. Furthermore, the inflorescence is not nodding at the apex as in the English Bluebell. When grown together the two species readily hybridize; intermediate plants are thus common in gardens.

Colour variations are found in *Hyacinthoides non-scripta*, and pink and white forms often occur both in the wild and in gardens.

Scilla non Scripta

Scilla peruviana Linnaeus

The confusing epithet which this Mediterranean plant carries can be traced back to the turn of the sixteenth century when Carolus Clusius is reported to have made a journey to Bristol to see it growing. He gave it the name *peruvianus* because he was apparently informed that the plant came from Peru, or possibly from *The Peru*, a ship which brought the bulb into England. Only shortly afterwards John Parkinson recognized the species as a Spanish plant, calling it the 'Great spanith starry Jacinth'. However, the previous name stuck and the first binomial published after 1753, and therefore the one we must accept under the International Rules of Nomenclature, was *Scilla peruviana.*

The species has a fairly wide distribution in the western Mediterranean—in Spain, Portugal, Algeria, Sicily, southern Italy and Sardinia. Not surprisingly therefore, it varies a great deal and has a wide range of variants, from few-flowered dwarfs to huge robust plants bearing up to one hundred flowers in a dense pyramid. Separate names exist for several of the more remarkable variants: for example, *Scilla cupanii*, for the short slender plant with up to ten flowers from Sicily; *Scilla hughii*, for the wide-leaved stout form from Isole Egadi; *Scilla elongata*, for a robust plant from Sardinia and Sicily, with densely ciliate leaves and bracts; and *Scilla sicula*, for one with paler flowers and sparsely ciliate leaves. However, intermediates occur and it is doubtful if these can justifiably be recognized as distinct species, interesting and 'different' as they are from a gardener's point of view.

Scilla peruviana is a fine vigorous summer-flowering bulb with a striking conical head of steely violet flowers, and there is also an equally attractive white form in cultivation.

1. Perianth segment and stamen 2. Pistil

Scilla Peruviana

LILIACEAE

Kniphofia linearifolia Baker

figured as *Tritoma uvaria*

REDHOT POKER

The Redhot Pokers were at one time included in the genus *Aloe* and it must be admitted that there is a vague overall similarity in the flower shape and colour, but there the relationship ends. Other generic names that have at times embraced the plants we now know as *Kniphofia* include *Aletris*, *Veltheimia* and *Tritoma*. However, it is Conrad Moench's name which takes priority over all others, and he called the genus after Johannes Hieronymus Kniphof, professor of medicine at Erfurt University during the early eighteenth century. Kniphof's *Herbarium Vivum* of 1762 was a most interesting publication in which 1,200 plant illustrations were reproduced by coating dried specimens with ink and pressing them on paper. The results were extremely good, and one illustration is of the plant now known as *Kniphofia uvaria*. However, Redouté's illustration does not show this species. According to Dr L.E. Codd, in *The South African Species of Kniphofia* (1968), this plate figures *Kniphofia linearifolia*. The two species have been much confused by botanists, but Codd considers that they can be distinguished in part by their leaves, which in *Kniphofia uvaria* are distinctly grey-green, very tough and fibrous and less than 1.5 centimetres wide, while in *Kniphofia linearifolia* they are rather more green than glaucous, up to 3 centimetres wide and softer in texture. In general *Kniphofia linearifolia* is a more robust plant with larger inflorescences. The majority of the Redhot Pokers, nearly fifty species in all, are natives of Africa, from South Africa to Ethiopia, with a few outlying ones in Madagascar and Arabia.

Kniphofia linearifolia occurs in the eastern Cape region, the Orange Free State, Natal and Transvaal and northwards into Zimbabwe. It is a mountain plant growing at altitudes of up to about 2,200 metres in dampish grassland.

Tritoma Uvaria

Tulipa acuminata Vahl ex Hornemann
figured as *Tulipa cornuta*

This curious tulip is not a wild species but a garden cultivar of some antiquity and, as with so many of these man-made hybrids or selections, it is impossible to come to a conclusion about its origin. It is quite likely that the much-narrowed perianth segments and strange colorations are caused by a virus disease, since these diseases are common in tulips and give rise to other deformities such as 'broken' colours and lacerated perianth segments. Some of these variants or mutations, increased by vegetative means, have been in cultivation for several centuries and a few have even become naturalized, thus giving the impression of being true species. For this reason, many of them have in fact been given specific names. They have been classified in a group referred to as the Neo-Tulipae.

Tulipa acuminata exists in both yellow and red forms; the two are shown in Redouté's illustration, and it seems that his plants were identical to those still available from nurseries under this name. Although far from beautiful, it has a certain charm about it and the fact that the stocks have been perpetuated for so long suggests that its appeal is fairly widespread; few nurserymen would bother with a freak if it had no commercial possibilities.

It is interesting to note that during the Ottoman period in Turkey the popular type of tulip grown had rather narrow long-pointed perianth segments and was raised in many colour forms. Indeed, there was such a cult surrounding the 'Istanbul Tulip' that committees were held to judge any new varieties and coloured illustrations of these still exist in the library of the Topkapi Museum. It is just possible that *Tulipa acuminata*, or plants very like it, date back to this period.

Tulipa Cornuta

LILIACEAE

Tulipa gesnerana Linnaeus variety *dracontia*

figured as *Tulipa gesneriana*

PARROT TULIP

Tulipa gesnerana is a group name which has been used for a wide range of 'garden tulips' of unknown origin, dating back to at least the sixteenth century when these showy plants were first brought into Europe. Already at this time, there were many varieties known to the Turks and it is probable that the origins of this 'species complex' lie in central Asia. However, through hybridization and selection the characters of the true wild species have become completely obscured and it is now impossible to even guess at their parentage.

The Ottoman Turks and Persians do not, however, appear to have had varieties such as the one illustrated here, in which the broken-coloured perianth segments were lacerated at the margins, and it seems that these were a product of European horticulture.

The classification of garden tulips was undertaken jointly in the twentieth century by the Royal General Dutch Bulb Grower's Society and the Royal Horticultural Society. At first they produced a list of twenty-three 'divisions', which were based largely on flower shape and flowering time or upon hybrids of a particular species, but they later reduced these to fifteen. The Parrot Tulips are one of these divisions and this plate clearly shows the group feature of fringed, or laciniate, segments. In this particular variety the flower is 'broken' by irregular splashes of red and yellow, but within this group there is a wide range of variants, from plain to bi-coloured in white, red, maroon, mauve, pink, orange or yellow. The name 'dragon' given to the form illustrated by Redouté seems eminently suitable for such a bizarre flower!

Bulb

Tulipa Gesneriana

LILIACEAE

Tulipa agenensis de Candolle
figured as *Tulipa oculus-solis*

RED TULIP OF BOLOGNA

This scarlet tulip was well known to the early herbalists, who called it the 'red tulip of Bologna' because it grew around that town, as well as in other places in Italy and southern France. It is however almost certainly naturalized in Europe and its origins probably lie much farther to the east, possibly in central Asia. Since this species is stoloniferous, it is very easy for it to become established by spreading vegetatively. Most of the large-flowered, rather gaudy species of *Tulipa* are native in Russian central Asia and Afghanistan, and the most likely explanation of the presence of *Tulipa agenensis* and several others in western Asia and Europe is that they were brought westwards by migrating peoples who later settled in Asia Minor. Tulips are mentioned in the writings of Omar Khayyam, and Seljuk artists adopted the flower as their emblem in the twelfth century. However, it seems that tulips were unknown in the gardens of Europe when the Flemish diplomat Busbecq, ambassador to Suleiman the Magnificent, wrote in 1554 that he had seen these gorgeous flowers in the gardens of Adrianople. Busbecq may well have been responsible for the introduction of tulips into Europe during the sixteenth century.

Tulipa agenensis is characterized by its bulb tunics with their thick felt of silky hairs on the inside, and its large red flowers which have a long narrow blackish blotch at the base of each perianth segment. There are several related species and Sir A. Daniel Hall, who made a thorough study of the genus, used the species figured here as the basis for his 'Sub-Section III, the Oculus-Solis Group'. The name *Tulipa oculus-solis* of J.F.B. de St-Amans cannot however be used for this species, since it is of a later date than *Tulipa agenensis*. There is another illustration of *Tulipa agenensis* in *Les Liliacées* (Vol. I, pl. 60).

Tulipa oculus-Solis

LILIACEAE

Veratrum nigrum Linnaeus

BLACK FALSE HELLEBORINE

This robust herbaceous perennial plant belongs to a genus of rhizomatous members of the lily family, the species of which are distributed widely in the northern temperate regions of Europe, Asia and North America. In Europe, however, there are only two rather variable species, one with small brownish or black flowers and one with much larger white or greenish-yellow flowers. The former, *Veratrum nigrum*, is the subject of this fine plate, but the latter, *Veratrum album* is the more impressive of the two. They can be distinguished not only by flower colour, but also by their leaves: the upper stem leaves are narrowly lanceolate in *Veratrum nigrum* and broadly lanceolate in *Veratrum album*. Moreover, the undersurface of the leaves is pubescent in the White Veratrum or White False Helleborine and glabrous in *Veratrum nigrum*, the Black False Helleborine as it is sometimes known. The ancient authors called them 'Hellebore', although they are in no way connected with the genus *Helleborus*, which is related to the *Ranunculus*.

Many of the species have been used locally for medicinal purposes and the root of the White Veratrum has been advocated as a sedative and as a cardio-vascular depressant. Dioscorides noted that it was useful as a mouse poison, so one would be well advised to treat *Veratrum* with great caution.

Veratrum nigrum is a plant of mountain meadows and light woodland in central, south and southeast Europe and eastwards through Russia to China. It and other species of the genus are unusual members of the lily family in having bisexual and unisexual flowers on one inflorescence, the upper ones often bearing stamens only.

Veratrum Nigrum

AGAVACEAE

Agave yuccifolia de Candolle
figured as *Agave yuccaefolia*

CENTURY PLANT

In this very large genus of about three hundred species, all are natives of the Americas, mainly Mexico, the southern United States and the West Indies. They have been included in the family Amaryllidaceae, but it is now generally accepted that a more realistic approach to their classification is to place them, together with some related genera such as *Furcraea* and *Yucca*, in a separate family, Agavaceae.

Agaves are mostly desert or semi-desert plants and their thick fleshy leaves with a leathery cuticle make them well adapted to overcoming long dry periods. Their symmetrical rosettes, usually stemless on the ground or with a short trunk, do not produce an inflorescence each year. Indeed, some of the species take many years to grow to flowering size and after flowering the rosette dies.

Although agaves are much cultivated around the world as ornamental plants, certain species have also been used as a source of fibre and soap and as a basis for two alcoholic drinks in Mexico, pulque and mescal.

The classification of this great number of species is not easy, for some of them flower very rarely and a few of the cultivated clones have never been recorded in flower. *Agave yuccifolia*, from Mexico, belongs to the subgenus *Littaea* in having a very dense spike-like inflorescence rather than a branching panicle such as the species of subgenus *Euagave* possess. Unlike many of the large agaves, the leaves are more or less unarmed on the margins, or at most have only a few soft prickles. The agaves are often called Century Plants, possibly because of the great age some of them reach before flowering.

1. Whole flower opened out 2. Pistil
3. Transverse section of ovary 4. Basal leaf (actual size)

Agave Yuccæfolia

AGAVACEAE

Furcraea foetida (Linnaeus) Haworth figured as *Furcroea gigantea*

The genus *Furcraea* was named in 1793 by Etienne Pierre Ventenat after Antoine François de Fourcroy, professor of chemistry at the Muséum National d'Histoire Naturelle in Paris and the author of *Leçons élémentaires d'histoire naturelle et de chimie*. Previously the species had been regarded as members of the genus *Agave* and indeed they are very similar. The large pendulous white flowers of *Furcraea foetida* do however look rather different from those of the agaves.

This huge plant has a long history of cultivation both in Europe and in the Mascarene Islands, and the excellent fibre obtained from its leaves is in fact known as Mauritius hemp. William Aiton records in *Hortus Kewensis* that it was cultivated at Hampton Court in 1690 and it was certainly grown in Europe nearly fifty years earlier than this. William Trelease, in his 'Observations on Furcraea' (*Annales du Jardin de Buitenzorg*, 1910) notes that it has been known in Mauritius since at least 1750 and that two forms occur there, one with prominently toothed leaves known as variety *willemetiana* and the typical form which is untoothed or with only a few teeth on the margin. The former has the local name of Aloës Creole and the latter Aloës Malgache.

It deserves the epithet *gigantea*, for its leaves can reach 2.5 metres in length and the inflorescence 9 metres. The earlier and therefore correct name of *Furcraea foetida* probably refers to the pungent scent of the flowers, a little reminiscent of soap. It is not surprising that in suitable habitats the plant has become semi-naturalized, for after it has flowered, a large number of bulbils form in the axils of the bracts of the inflorescence and on falling to the ground these make new plants.

Furcraea is a genus of perhaps as many as forty species, although some authorities recognize only a dozen or so. They are mainly from Colombia, Venezuela and Brazil with some species reaching the West Indies and Mexico. The species shown here is probably native in Brazil although it is also recorded as occurring in Central America.

Furcroea Gigantea

AGAVACEAE

Polianthes tuberosa Linnaeus

TUBEROSE

Although this is unknown as a wild plant it is reasonable to assume that it is of Mexican origin for the whole genus, comprising about ten species, occurs there. *Polianthes tuberosa* bears some similarity to the smaller and more slender-leaved *Polianthes gracilis*. J.N. Rose, who published a revision of the whole genus in 1903, felt inclined to treat this latter as the wild form from which the more robust *Polianthes tuberosa* arose.

There is a long and detailed account of the cultivation and origins of the Tuberose by R.A. Salisbury in the *Transactions of the Horticultural Society of London* in 1812 and mention is made of a double form that was known in 1731. This is still grown, probably more widely than the more attractive single form which is figured here by Redouté. As early as 1594, however, Carolus Clusius received a specimen of the single form; this was sent to him by Paludanus, a physician in Rome who had in turn received it from Simon de Tovar in Seville. It seems highly probable that the spread of this beautiful plant followed the Spanish exploitation of Mexico. There is for instance an early record of it being introduced into Luzon by the Spaniards and this may account for the East Indies being cited in many older works as the country of origin of *Polianthes*. Living material may well have been imported into Europe from the East since it has been cultivated there for its intensely fragrant flowers for many centuries.

At one time the cultivation of the Tuberose was considered a worthwhile occupation, for Salisbury enters into the economics of the crop in some detail. He stated that in a quarter-acre plot it was possible to grow 15,125 of the tuberous roots, which at 3d. each brought in £189 1s. 6d. and even after deduction of production costs 'the profits of a bed of the Tuberoses would be considerable'. The methods used however were laborious and the present cultivation in Britain is so drastically reduced that it is rarely seen, although the roots are readily obtainable through the Dutch bulb trade.

1. Flower opened out to show stamens 2. Pistil

Polianthes Tuberosa

AGAVACEAE

Yucca gloriosa Linnaeus

The yuccas are a most interesting group of plants. Like *Agave* and *Furcraea*, which belong to the same family, they have tough fibres in their leaves and these have been used as fibre to make yarn, although they are rather too short and coarse to be of great commercial value. These striking plants, numbering about forty species, all occur in the southern United States, Mexico and the West Indies and have been very extensively used by local people. For example, the roots possess saponifying properties and were at one time used to prepare soap, the leaves make a serviceable thatch for houses, the flowers can be eaten raw or cooked, the seeds have been thought effective in the treatment of dysentery and the fleshy fruits of some species are edible and can also be fermented and distilled to provide an alcoholic drink. In addition, yuccas are of course very ornamental and are planted in gardens all over the world both in temperate and in tropical regions. They are particularly useful in dry places, as they can resist arid conditions well.

Yucca gloriosa is an impressive species even when not in flower, with its dense tufts of one hundred or more leaves at the apex of a short stem. The leaf margins are smooth, not spiny or furnished with long loose threads as they are in many of the species. If, as is almost certain, this is the same plant that was known to both John Gerard and John Parkinson in the late sixteenth and early seventeenth centuries, we can be sure that it has been in cultivation for at least three hundred and fifty years, probably over four hundred. It occurs in the coastal regions of Carolina and Georgia in marshy places and among sand dunes in moist depressions.

1. Leaf 2. Inflorescence 3. Stem

Yucca Gloriosa

ALISMATACEAE

Sagittaria lancifolia Linnaeus

figured as *Sagittaria ovata*

ARROWHEAD

The Arrowheads are a worldwide group of plants, although most of the twenty or so species are American. They belong to the family Alismataceae which bears a great superficial similarity to the Ranunculus family although there is probably no real link between the two, the Alismataceae being in fact true monocotyledons while the latter are dicotyledons. The main resemblances are in the cluster of many stamens and the dense heads of fruits called achenes.

Sagittarias are all water plants, usually growing near the margins of fresh-water swamps and streams. They are capable of producing different leaves depending upon whether these are submerged, floating or aerial: the underwater ones are ribbon-like, the floating ones broad and often ovate and the exposed ones usually arrow-shaped. The species depicted here however has lance-like aerial leaves which are sometimes broader and nearly elliptical in shape, but without the basal lobes which give the arrow-like shape. The name *sagitta* is Latin for an arrow.

Sagittaria lancifolia is a native of South and Central America, the southern United States and the West Indies. It is a robust and showy species growing 40 to 200 centimetres in height with numerous whorls of flowers up the stem. These give the appearance of being three-petalled although there are two whorls of three perianth segments – the outer ones smaller and greenish and looking like sepals. The lower flowers of the inflorescence are usually pistillate, that is female only, while the upper ones are usually staminate, or male. Plants with this structure are described as monoecious, while some of the other species are dioecious, having the male and female flowers on separate plants.

An alternative name for some species of *Sagittaria* is Duck Potato because the swollen tubers on the rootstock are eaten by certain water birds. These tubers are rich in starch and have also been used as food by some American Indian tribes.

Sagittaria Ovata

ALSTROEMERIACEAE

Alstroemeria ligtu Linnaeus

The alstroemerias are a showy group of plants occurring entirely in South America and comprising perhaps as many as fifty species. Most of these are unknown to horticulturists in Europe, but a few species have long been cultivated as herbaceous border plants and for cut flowers. Together with three other even less well known genera, they form the family Alstroemeriaceae, although they have been included with the Amaryllidaceae by some authorities. The features which make them immediately distinguishable from the latter family are the rhizomatous fleshy rootstocks, the twisted stem leaves and racemose or umbellate inflorescences. In the Amaryllidaceae the underground part is normally bulbous, the leaves usually basal and not twisted and the flowers are produced in an umbel on a leafless scape. The leaf character of *Alstroemeria* is curious, for it is often the stalk, or at least the very base of the leaf, which is twisted through 180° so that the true upper surface faces downwards.

Alstroemeria ligtu is one of the better known species and has been cultivated in European gardens for at least two centuries. It has been used as one of the parents of a range of colourful hybrids that are often referred to collectively as the 'ligtu hybrids'. The wild species, which takes its specific name from the local vernacular name in the Andes, has flowers of a predominantly pinkish or reddish colour sometimes marked on the upper perianth segments with red spots and lines on a yellow ground and with a brownish point at the apex. The upper segments are usually white in the lower portion, as in Redouté's illustration. There is, according to A. Garaventa, who has studied the Chilean species of *Alstroemeria*, a yellow form which occurs in the same areas as the more usual reddish or pinkish variety.

The genus is named after the Swedish botanist Clas Alströmer (1736–1794).

1. Perianth segment 2. Pistil and stamens

Alstroemeria Ligtu

AMARYLLIDACEAE

Sprekelia formosissima (Linnaeus) Herbert
figured as *Amaryllis formosissima*

JACOBEAN LILY

The Aztec Lily, Jacobean Lily or St James's Lily is a native of Mexico and Guatemala, growing at altitudes of up to 2,500 metres in rocky hills, on cliffs and on lightly wooded slopes. It was formerly placed in the genus *Amaryllis* but is now considered to be the only species in the genus *Sprekelia*, which is named in honour of J.H. von Sprekelsen who sent the plant to Linnaeus. It differs from both *Amaryllis* and the related *Hippeastrum* in its very irregular flower. The common name St James's Lily was inspired by the arrangement and colour of the perianth segments which were thought to resemble the cross embroidered on the mantles of the Spanish Knights of St James. James is a variant of Jacob, hence the alternative name.

Sprekelia formosissima is thought to have been introduced to cultivation in Europe in 1593 and it has been widely grown since then as a pot plant, although it is nearly hardy in Britain and can be grown outside in Mediterranean climates. The flowering period is variable and the gorgeous solitary flowers can be produced at any time from spring to autumn. There are several slight colour variants and some of these have been given names. The best known one is variety *karwinskii* which has paler segments, margined with white.

Whether cultivated under glass as a house plant or in the open garden, *Sprekelia* bulbs must be kept dry and away from frost during the winter months when they are dormant. Propagation is achieved by detaching the young bulblets which are produced around the parent bulb.

Pistil

Amaryllis Formosissima

Hippeastrum vittatum (L'Héritier) Herbert
figured as *Amaryllis vittata*
AMARYLLIS

Although known as 'amaryllis', this well-known and large-flowered plant is generally considered to belong to the South American genus *Hippeastrum*, whereas the true *Amaryllis belladonna* is a South African plant, a solitary species more closely related to *Crinum*.

Hippeastrum vittatum is a plant with a confused history. It was first thought to be South African in origin but is now known to occur wild in the Andes of Peru. Since its introduction into European gardens in the mid-eighteenth century it has been a popular greenhouse or conservatory plant, very showy and rather easy to grow. It has also been used as one of the parents of a range of hybrids and there are now so many that it is difficult to say which species are present in any one cultivar.

The original wild species is, as its name suggests, striped with eye-catching reddish-purple bands on a white ground. Even before the end of the eighteenth century, Mr Johnson, an English watchmaker from Prescot, crossed it with the red *Hippeastrum reginae* to produce *Hippeastrum* × *Johnsonii*. By further hybridization intermediates followed this; these have bright red flowers with a greenish base and white streaks on the perianth segments.

Hippeastrum is a large genus of around fifty species or more, entirely confined to Central and South America. They can be divided into groups or subgenera according to the length of the perianth tube and whether or not the throat of the flower is constricted by a 'collar' which nearly closes it. *Hippeastrum vittatum* has a short tube and the throat is not constricted, so it falls into J.G. Baker's subgenus *Lais*.

This, together with many other species and hybrids, is a popular and showy pot plant but, although noted by Lt-Col. C.H. Grey in his *Hardy Bulbs* as being nearly hardy in the south of England, it can be really successfully grown only in a greenhouse or as a house plant.

Pistil

Amaryllis Vittata

AMARYLLIDACEAE

Crinum erubescens Aiton

The large genus *Crinum* is distributed almost throughout the tropics of the world and is one that is difficult to classify. There are undoubtedly many species, but the differences between them are often not at all obvious and it is only after gaining considerable familiarity with the living plants and their ecology that one may distinguish the features that separate them.

Fortunately the species fall into three fairly distinct groups, classified as subgenera, and this makes the genus rather more manageable. *Crinum erubescens*, which occurs in the wild in Central and South America, is in the subgenus *Platyaster*. This group is characterized by having flowers which are more or less erect and acutely tipped perianth segments which are about 1.5 centimetres or more wide. Subgenus *Stenaster* also has upright flowers but the perianth segments are bluntly tipped and 1 centimetre or less wide, while subgenus *Codonocrinum* has funnel-shaped flowers with broad-segments and a curved tube which usually gives the blooms a pendulous aspect.

Crinum erubescens is a variable plant and has many forms that differ slightly from the one shown in this plate. It normally has about ten to fifteen strap-like thin-textured leaves 5 to 8 centimetres wide and 50 to 90 centimetres long, furnished with fine teeth on the margins. Usually they slightly overtop the flower stem which carries an umbel of four to twelve delicately scented flowers.

Crinum Erubescens

Crinum jagus (Thompson) Dandy
figured as *Crinum giganteum*

The genus *Crinum* contains a great many very beautiful species – in fact over one hundred and fifty have been described from the tropical and warm temperate regions of Africa, Asia and America. Only a handful of these, however, have found their way into gardens and all but one or two are quite unknown to the majority of horticulturists.

Many of the species have either distinctly funnel-shaped flowers or flowers which are somewhat star-shaped, with narrow spreading perianth segments. The West African *Crinum jagus* differs from both these groups in having very broadly bell-shaped or even flattish flowers with wide segments which are unstriped, unlike several of the African species whose segments have each a pink or reddish stripe along the centre. In flower shape *Crinum jagus* resembles the South African *Crinum moorei*, although the latter usually has pink-tinged flowers and rather narrower leaves than the former. *Crinum moorei*, which inhabits damp grassy places in the eastern Cape, is hardy in gardens as far north as southern Britain, but its West African relative is most definitely a warm greenhouse plant which can also be grown most successfully in a container in a heated conservatory. The umbels of large white flowers are produced in summer and have a delicious vanilla-like scent.

In the wild *Crinum jagus* inhabits forest swamps in Sierra Leone, Liberia, Ivory Coast, Ghana and Nigeria.

The epithet *jagus*, validly published by Thompson, is neither Latin nor Greek but, as R.A. Salisbury revealed in his posthumous *The Genera of Plants*, p. 123 (1866), owes its origin to the blunder of a Scottish gardener employed by Lee and Kennedy 'transforming by his Northern pronunciation *Gigas* into *Jagus*, in which latter way the work is printed'. According to Salisbury, 'a little before it was first figured by Thompson in 1799 I had proposed the name of *Gigas* . . . and this *Scotch* gardener, happening to be present in the Stove [hothouse] at the christening' not only misheard but applied it to a different species, the present one from West Africa and not the gigantic Australian species intended by Salisbury.

Crinum Giganteum

AMARYLLIDACEAE

Galanthus nivalis Linnaeus

SNOWDROP

It has been suggested that the omission of the Snowdrop from Shakespeare's writings can be attributed to the lengthy cumbersome names which this well-loved plant possessed at the time. In the sixteenth and seventeenth centuries John Gerard and John Parkinson referred to it as the 'Timely flowering bulbous violet' or 'Lesser early bulbous violet', not the most euphonious of names. Indeed, the earliest mention of the name Snowdrop seems to have been in Robert Boyle's *Colours* of 1664. Even later came the Latin name, *Galanthus*, which Linnaeus created in his *Systema Naturae* of 1735. Before then *Galanthus nivalis* was either included within the genus *Leucojum*, now restricted to the related Snowflakes, or was treated as a *Narcissus.*

Some of the English common names by which the plant used to be known are very beautiful; for example Fair Maids of February and Candlemas Bells. Both of these are connected with the Feast of Purification on 2 February when there were processions of maidens in white robes.

As with several other popular garden plants that have been in cultivation for centuries, there are now a host of variants of *Galanthus nivalis*, many with cultivar names. These come in a great range of subtle shapes and sizes of flower and some (e.g., 'Scharlokii') have a long bisected spathe overtopping the flower like a pair of ears. Some variants (e.g., 'Lutescens') have yellow rather than green markings on the inner segments and some (e.g., 'Poculiformis') have the inner and outer segments almost equal in size. Particularly attractive are for example 'Viridapicis', in which there are green marks at the apex of the three larger segments as well as the inner smaller ones; and in some forms (e.g., 'Virescens') the whole flower takes on a greenish shade. Most of these variants are easily increased by bulb division, so it is not surprising that they have persisted in cultivation for many years.

In the wild, *Galanthus nivalis* is a woodland plant, distributed throughout Europe over a wide altitude range from sea level up to about 2,700 metres.

Galanthus Nivalis

AMARYLLIDACEAE

Scadoxus multiflorus (Martyn) Rafinesque
figured as *Haemanthus multiflorus*

FIREBALL LILY

The tropical African Fireball Lily has for a great many years been placed in the genus *Haemanthus*, together with a number of other similar species. Recently, however, studies by V. & D. Müller-Doblies, I. Nørdal and I. Friis have shown that it and its relatives should be separated as a distinct genus, the oldest name for which is *Scadoxus* of C.S. Rafinesque (*Flora Telluriana*, 1838). Prior to this reinstatement of *Scadoxus* the genus *Haemanthus*, comprising some twenty species, had been divided into two subgenera: subgenus *Haemanthus*, restricted to South Africa, and subgenus *Nerissa* (now genus *Scadoxus*) which is mainly tropical African but just extends southwards into Natal and the eastern Cape. The two subgenera differ not only in morphological characters but also in their ecology and it is suggested by the authors mentioned above that they represent evolutionary lines which were separated at a very early stage in the development.

The early history of the African Fireball Lily as far as botanists are concerned is a confused one, for it was probably first noted and illustrated rather poorly in 1608, as an orchid, *Satyrium*. The first valid name following the International Rules of Nomenclature was given by Th. Martyn who published a small paper on the species *Haemanthus multiflorus* in 1795, together with a beautiful illustration by F.P. Nodder. It was said to have been introduced from Sierra Leone, possibly by Adam Afzelius.

A widespread plant in a wide range of habitats throughout tropical Africa, it varies considerably over this area, especially in size and number of flowers in the umbel. In fact over thirty 'species' have been described within what is now recognized as *Scadoxus multiflorus*.

The illustration by Redouté shows a form which has a less spherical umbel of flowers than is normal for the species.

1. Flower opened out 2. Pistil

Hæmanthus Multiflorus

Narcissus pseudonarcissus Linnaeus subspecies *moschatus* (Linnaeus) Baker

figured as *Narcissus candidissimus*

This small white trumpet daffodil is one of the more distinctive members of a group of *Narcissus* species and their variants which were known to de Candolle as the Pseudonarcissus section. There have been many widely differing approaches to the problem of classifying this complex, ranging from that of H.W. Pugsley, who recognized twenty-seven species (*Journal of the Royal Horticultural Society*, Volume 58, 1933) to that of A. Fernandes who, at the other extreme, gave specific status to only four. The most recent view, taken by Professor D.A. Webb in *Flora Europaea*, Volume 5 (1980), is that in Europe there are in this section six species, one of which is the highly variable *Narcissus pseudonarcissus*. This species is then further divided into seven subspecies, based on the flower colour, the shape of the corona or trumpet, and the length of the pedicel. Subspecies *moschatus*, which includes as a synonym the well-known *Narcissus alpestris* of Pugsley, is wholly white-flowered and therefore immediately distinguishable from the other subspecies, in which the corona and perianth segment are varying shades of cream or yellow.

There are, however, a few features other than colour which make it distinctive. The flower is somewhat drooping and the twisted perianth segments, instead of standing out horizontally as they do in most daffodils, are held almost alongside the trumpet, so the whole flower has a somewhat wilted appearance. The corona usually does not flare outwards at the mouth as much as it does in most of the garden trumpet daffodils.

This attractive variant of *Narcissus pseudonarcissus* occurs wild in the Pyrenees and the Cantabrian mountains and has probably been in cultivation for about four hundred years.

1. Capsule cut to show seeds 2. A single seed

Narcissus Candidissimus

AMARYLLIDACEAE

Narcissus pseudonarcissus Linnaeus

WILD DAFFODIL, LENT LILY

This most familiar of all the wild daffodil species is the most difficult when it comes to deciding the botanical status of its many variants. Whether one accepts H.W. Pugsley's recognition of twenty-seven different species within de Candolle's section Pseudonarcissus, or the view of A. Fernandes according to which there are only four species in the section, the fact seems to be that there is almost continuous variation in the plants of this group, and no really clear-cut divisions.

The plant illustrated here shows features which suggest that it is near to 'typical' *Narcissus pseudonarcissus*, although the very flared mouth of the trumpet is reminiscent of *Narcissus pallidiflorus.*

The Lent Lily is a widespread wild plant in western Europe, from Portugal, Spain and France northwards to Britain and Holland. It is naturalized in several other countries and has been much used as one of the parents of a great range of trumpet daffodils.

1. Perianth opened out to show stamens 2. Pistil

Narcissus Pseudo-narcissus

Narcissus tazetta Linnaeus

TAZETTA NARCISSUS

This well-known cluster-headed narcissus has, like several other species, been in cultivation for many centuries and it is now difficult to be sure of its natural wild distribution. It is found growing in meadows and cultivated areas such as olive groves in the Mediterranean region from Spain, Portugal and North Africa eastwards to Turkey. It is probably native in these areas, but farther east in Iran, Kashmir, China and Japan it is almost certainly naturalized. The unusual feature of this distribution is that the species appears to have been spread eastwards by man whereas most of our delightful spring bulbous plants such as tulips and hyacinths reached Europe from the Orient. Since for the most part the genus *Narcissus* is centred on the western Mediterranean, it seems reasonable to assume from the available facts that the Tazetta Narcissus is also a European plant, perhaps extending naturally to western Asia Minor. If this is so, the dispersal of the species must have taken place in very early times, for the Chinese were depicting it in single and double forms in the thirteenth century.

There are many variants of *Narcissus tazetta* and over the years many 'species' have been described, differing mainly in the size and colour of the corona and perianth segments, the colour of the leaves and the cross-sectional shape of the stem. It seems best to regard these as subspecies of one widespread and very variable species (see *Flora Europaea*, Volume 5, 1980).

The major characteristics of *Narcissus tazetta* are the several-flowered umbel and fairly shallow cup or corona, which is yellow or orange as distinct from the white cup of *Narcissus papyraceus*. The perianth can be white or yellow and it is possible to recognize three subspecies, *tazetta, aureus* and *italicus*, from the colour of their perianths—all three have a yellow cup, but *tazetta* has a white perianth, *aureus* has a yellow one and *italicus* a creamy one.

The flowers of all the variants of Tazetta are very sweetly scented and it is so popular as a cut flower that many forms of it are grown in mild districts and transported in winter to more northerly markets.

Perianth opened out

Narcissus Tazetta

AMARYLLIDACEAE

Hymenocallis rotata (Ker-Gawler) Herbert
figured as *Pancratium disciforme*

SPIDER LILY

The genus *Hymenocallis* is so called because of the membrane-like 'cup' in the centre of the flower. Formerly, the species that belong to this wholly New World genus were included in *Pancratium*, its Old World equivalent. J.R. Sealy in the *Kew Bulletin* of 1954 reviewed the species and showed that *Hymenocallis rotata* and *Hymenocallis disciforme* were in fact synonyms so that the latter epithet, which was provided by de Candolle for Redouté's illustration, must give way to the earlier one provided by John Bellenden Ker-Gawler.

The Spider Lilies number about forty to fifty species, distributed from the warmer states of North America through Mexico into the tropical and subtropical regions of South America. The species illustrated here occurs only in the southwestern United States and is possibly also a native of Jamaica since there is one old record of its occurrence there. Its typical habitat is in wet grassy or marshy places.

The taxonomy of *Hymenocallis* is, it seems, not well understood and J.R. Sealy acknowledged that there was still a great deal of research and field observation required in the genus. The species are separated largely by the shape and size of the leaves, the length of the perianth tube and filaments and the relative measurements of the staminal cup. *Hymenocallis rotata* is characterized by its blunt strap-shaped leaves, 3 centimetres or less wide, and by the large cup which is tubular at the base and flaring widely at the apex. The perianth segments spread out widely from the base whereas in some species they tend to lie alongside the more bowl-shaped staminal cup.

Although not widely cultivated at any time, *Hymenocallis rotata* has a long history in gardens, having been introduced into Europe by A. Michaux in 1790. The genus, however, was known as early as 1535 (although not as *Hymenocallis*) to Oviedo, who illustrated a species from Panama in his *Historia natural y general de las Indias*.

Pancratium Disciforme

AMARYLLIDACEAE

Ananas comosus (Linnaeus) Merrill
figured as *Bromelia ananas*

PINEAPPLE

The Pineapple belongs to the almost entirely South American family Bromeliaceae, a very large and characteristic group of nearly fourteen hundred species. Many are grown as ornamental plants for their flowers or foliage and some for their curiosity value. Although the Bromeliaceae are almost all epiphytes, *Ananas* is one of the exceptions, being a ground-dweller. A lot of bromeliads have channelled leaves which form a rosette that catches water in the funnel-shaped centre formed by the leaf bases. The hairs in the funnel absorb the water in the same way as do the root hairs in terrestrial plants.

Ananas comosus belongs to a South American genus of only five species and is grouped in Hermann A.T. Harms's classification in the sub-family Bromelioideae on account of its toothed leaves, berry-like fruit, inferior ovary and non-hairy seeds. The fleshy fruit is of course man's main interest in the Pineapple. It is an important crop in Hawaii and Malaysia and is also produced commercially, on a smaller scale, in several other tropical countries. The fruit has an interesting structure, for it is the product of a whole spike of a hundred or more flowers, the ovaries of which, after fertilization, become fused together and expand into one large fruit, the well-known pineapple. The central axis of the spike continues to develop at its apex into a tuft of leaves, forming the crown. This crown can be removed and planted as a cutting to produce a new plant, or the sucker-like growths at the base of the rosettes can be used for propagation purposes. Because of the very spiny-margined leaves, pineapples are difficult to handle and the pickers need the protection of strong gloves and, since the plants are nearly stemless on the ground, special trousers to protect their legs. Plant breeders have selected smooth-leaved varieties, but these are not yet widely planted.

The deliciously flavoured Pineapple fruit is very rich in sugar and has a moderate vitamin A and C content. It is used extensively as a fresh fruit and is also canned in chunks or slices.

Bromelia Ananas

AMARYLLIDACEAE

Pitcairnia bifrons (Lindley) R.W. Read figured as *Pitcairnia latifolia*

The genus *Pitcairnia*, belonging to the Bromeliaceae, was named in 1788 by Charles Louis L'Héritier in honour of the physician William Pitcairn (1711–1791) who built up a botanist's garden in Islington. It is in no way connected with Pitcairn Island of 'Mutiny on the Bounty' fame. *Pitcairnia* species are natives of a quite different region, mainly in tropical South America and the West Indies with one outlying species in West Africa. This distribution is most interesting, for it is yet another factor to be considered in the 'continental drift' theory of present-day land mass positions and the relationships between the continents.

There are about two hundred and fifty species of *Pitcairnia* in all, widely distributed in the warmer parts of South America. However, the species illustrated here is confined to the Lesser Antilles. Within this island group it has been recorded in St Christopher (St Kitts) and Guadeloupe in the Leeward Islands, and Martinique in the Windward Islands. It is a terrestrial plant, unlike many bromeliads which are epiphytic, and it inhabits grassy areas and rocky summits anywhere between sea level and 1,500 metres.

Pitcairnia bifrons is a robust plant with smooth-margined leaves 70 centimetres or more in length and an inflorescence about the same height with a simple raceme of red, orange or yellow flowers up to 30 centimetres long. The silvery flowering stem is an attractive foil for the brightly coloured flowers and although there are many much more striking members of this large family, it is nevertheless a showy plant and has been cultivated for nearly two centuries in Europe. It of course bears very little resemblance to the Pineapple, which is also in the Bromeliaceae (see page 143), and is placed by botanists in a quite separate group of the family, namely the Pitcairnioideae. The fruits of *Pitcairnia* are dry capsules, quite unlike the well-known fleshy delicious fruits of *Ananas*.

1–2. Whole flower 3. Perianth segment showing nectary
4. Pistil and stamens 5. Single stamens 6. Pistil
7. Tip of style showing stigmas
8. Transverse section of unripe fruit
9. Unripe fruit protected by calyx

Pitcairnia Latifolia

BUTOMACEAE

Butomus umbellatus Linnaeus

FLOWERING RUSH

The initial statement accompanying the original plate in Volume IV of *Les Liliacées*, 'Le Butome est très facile à reconnaître' (*Butomus* is very easy to recognize) is indeed true, for there is really no other plant like it. *Butomus umbellatus* is the only species in the genus, and is in fact considered by some authorities to be the only species in the family Butomaceae.

The Flowering Rush is a widespread plant throughout Europe and temperate Asia, inhabiting still or slow water, where its thick rootstocks creep along the margins of ponds, ditches, canals and rivers. It is also present in North America, on the St Lawrence and Great Lakes, but here it has escaped from cultivation. In Europe it is undoubtedly one of the most handsome of all water plants and has of course a long history of cultivation, being as at home in a garden pond as in its wild habitat.

The name *Butomus* is apparently derived from two Greek words, *bous*, an ox and *temno*, cut, because the leaves are sharp-edged and are said to hurt the mouths of browsing cattle.

Redouté's illustration shows a very pale pinkish-flowered form, but *Butomus* can vary considerably in colour, from nearly white to an attractive rose pink. The six perianth segments are arranged in two whorls, the outer three being small and green-tinged and the inner three larger and coloured. The flower is made more conspicuous by the nine reddish stamens, a very unusual number in the monocotyledons. These behave curiously in that six of them produce pollen and turn outwards on curving filaments while the other three mature slightly later and remain erect so that their pollen is in close proximity to the stigmas. It appears therefore that the first six to mature serve for cross-pollination and the others are a self-pollinating system as insurance in case all else fails.

In addition to the beauty of the many-flowered umbels, the Flowering Rush has a delicate fragrance.

Butomus Umbellatus

CANNACEAE

Canna latifolia Miller
figured as *Canna gigantea*

CANNA

Cannas, belonging to the family Cannaceae, are loosely related to the gingers (Zingiberaceae), the bananas (Musaceae) and to Marantaceae. At times these have all been merged in one large group known as the Scitamineae but it is now generally accepted that in reality there are several distinct families involved. *Canna*, like the other genera belonging to these families, has very irregular flowers, the parts of which need some explanation as they are rather complex. They consist of six perianth segments—three small, greenish and sepal-like ones and three larger, coloured ones which are usually narrow and pointed. The most prominent part of the flower is made up of stamens which have become sterile, enlarged and petal-like and usually highly coloured. One of these is often reflexed to form a lip. Only one fertile pollen-producing stamen remains and there is a single style. In *Canna latifolia* the green sepals are very unequal, the scarlet petals have long points and are about 5 centimetres in length, and the brick-red sterile stamens, or staminodes as they are known, are fused into a tube at the base and are somewhat twisted in the expanded upper part. The synonym *Canna gigantea* refers to the fact that it is normally a robust plant, growing about 3 metres tall with ovate purple-tinged leaves a metre or more long.

Cannas, probably about twenty-five species in all, are native both in tropical America and Asia and are popular garden plants for their foliage and flowers. In the mid-nineteenth century hybridization of the species began, at first producing rather tall small-flowered cultivars; later, however, more compact larger-flowered hybrids were raised and these were called either French or Crozy Cannas. A race with rather more attractive flowers are known as Orchid-flowered Cannas, or Italian Cannas after their country of origin.

As garden plants they are not frost-hardy, so in temperate countries they are used either as summer bedding plants or for pot culture in greenhouses and conservatories. They are very easy to grow and may be increased by division of the stout rhizomes.

Canna latifolia is a native of South America, mainly Brazil and Colombia, where it grows in damp places at altitudes of up to 2,000 metres.

1. Inflorescence 2. Leaves 3. Stamen and style
4. Ovary and sepals 5. Ovary and lower part of style

Canna Gigantea

COMMELINACEAE

Commelina tuberosa Linnaeus

As a garden plant, *Commelina tuberosa* is often known by its synonym *Commelina coelestis*, an apt name describing the intense blue flowers. However, the correct epithet, *tuberosa*, is not at all misleading, for the roots consist of a cluster of thick fleshy tubers. These give rise in summer to leafy stems up to 60 centimetres in height; and from July to September a mass of the flowers is produced in succession, although each one lasts only a few hours and is withered by about mid-afternoon. A good-sized clump of this plant is a very dramatic sight when in full flower, with masses of the delicate flowers giving the appearance of hovering blue butterflies. It is a native of Mexico and often occurs wild in the mountains at altitudes around 1,500 to 2,500 metres. Being thus accustomed to cool conditions, it is hardy in areas such as southern England. It has long been known as a garden plant and there are records of its cultivation dating from as early as 1732.

Commelina tuberosa belongs to a large genus of about two hundred and thirty species; these are mainly tropical and subtropical and are distributed throughout the world. In the family Commelinaceae this genus is one of the easiest to recognize because its flowers are produced in succession from a curious boat-like spathe—in the case of the species illustrated here a very large one—sometimes as much as 4 centimetres in length, with a long-pointed apex. A great many *Commelina* species have blue flowers and they usually have two of the petals much larger than the rest and standing up like a pair of ears, but *Commelina tuberosa* is distinctive in having three equal petals. White-flowered and variegated forms have been selected for cultivation.

Commelina Tuberosa

Rhoeo spathacea (Swartz) Stearn
figured as *Tradescantia discolor*
OYSTER PLANT, BOAT LILY

The Oyster Plant or Oyster Lily is frequently cultivated in many parts of the world, both in gardens of the tropics and as a greenhouse or house plant in colder climates. In many tropical countries it has become naturalized in semi-wild situations although in its natural state it is confined to Central America.

A very distinctive plant, it need not be confused with any other member of the family Commelinaceae; hence its presently accepted position on its own in a separate genus. The tufts of long purple-tinged leaves make it instantly recognizable, even without flowers, but the boat-shaped spathes which enclose the white flowers add to the overall effect of uniqueness, and have given rise to the alternative vernacular name of Ladies in a Boat.

Unlike many of the Commelinaceae, *Rhoeo spathacea* has regularly shaped flowers with three equal outer and three equal inner perianth segments, and six equal fertile stamens. Most members of the family have unequal segments or markedly different stamens, some of them converted into curiously shaped sterile 'staminodes'. In common with a lot of the Commelinaceae, the individual flowers of *Rhoeo spathacea* are short-lived, but a long succession is produced from each boat-like set of spathes.

Rhoeo spathacea is apparently native in British Honduras and Guatemala, but it is so well established in other neighbouring countries that it is difficult to be sure of the precise natural distribution.

Flower seen from below

Tradescantia Discolor

IRIDACEAE

Iris amoena de Candolle

The plant illustrated here and described fully in the original text by A.P. de Candolle is very different from the group of garden plants known by iris fanciers today as the Amoenas. The present definition of this group, quoted from the recent American Iris Society's publication *The World of Irises*, is 'a bearded Iris with white standards and colored falls'. The colour of the falls is not necessarily blue and it is possible now to have Amoenas with, for example, pink falls and white standards.

No plant exactly matching Redouté's *Iris amoena* is known in the wild and we may assume it to be a hybrid involving *Iris variegata*. As mentioned in the text on page 188, this eastern European species is extremely variable in colour and some of the forms have purplish veining on a white, rather than yellow, background. In its flower colour *Iris amoena* therefore resembles some forms of this species; but in its grey-green leaves it differs, for those of *Iris variegata* are normally quite green with no markedly glaucous appearance. There is, however, a rather ribbed aspect to the leaves, which is a feature of the Variegated Iris, so it seems fairly certain that herein lie the origins of *Iris amoena*.

1. Outer perianth segment
2. Ovary, style branches and stamen

Iris Amœna

Iris cristata Aiton

DWARF CRESTED IRIS

This beautiful miniature iris is placed in the same group of species as the oriental *Iris japonica*, namely the Evansia section of the genus. This may seem rather remarkable in view of the considerable differences in appearance between these two species, but they are brought together by one point in common: both possess a raised yellow portion, or crest, on the outermost three segments, or falls as they are more generally known. This is not in the form of a beard as in *Iris germanica* and other members of the bearded group, but rather a prominent ragged ridge like a cockscomb. Hence its local name of Dwarf Crested Iris in its native lands of the central-southern states of North America, where it is one of the most beautiful of the small spring flowers. Widespread from Maryland to Missouri and south to Oklahoma and Georgia, it tends to be a moisture-loving species of lowland woods.

Farther to the north, in Illinois, a near relative, *Iris lacustris*, occurs in gravelly or sandy places on the shores of Lake Huron and Lake Michigan. In fact the two are so similar that the northern plant is sometimes treated botanically as a variety of *Iris cristata*. Both are of extreme delicacy and from a gardener's point of view there is little to choose between them, although *Iris lacustris* is rather smaller with narrower leaves than its southern neighbour. There are extremely beautiful white forms of both, which retain the yellow crest in spite of being albinos.

1. Outer perianth segment 2. Ovary, style branches and stamen 3. Young fruit

Iris Cristata

Iris japonica Thunberg
figured as *Iris fimbriata*

This very elegant iris belongs to the group known as the Evansia irises, named by R.A. Salisbury in 1812 after Thomas Evans who introduced the first living material of the species into Britain. Since 1794, when C.P. Thunberg first described *Iris japonica*, several others have been added to the group, but it remains a small section of less than ten species, distinguished from other irises by the presence of a cockscomb-like crest on the three outer petals, or falls. Unlike many of the rhizomatous *Iris*, this species produces rather slender tough green rhizomes which send out long stolon-like shoots, rooting down at the apex and giving rise to further fans of glossy green leaves. With this habit of growth it can rapidly form extensive patches and is a most impressive sight when flowering freely. Unfortunately it does not always do so and appears to require a good warm summer in order to produce strong flowering-size fans.

Iris japonica is thought to be native in China and Japan, although one cannot be sure of the origins of a plant which is so easy to propagate and distribute.

The form illustrated here is of a slightly darker lavender than is usually seen, but there is a little variation in flower colour between different plants of *Iris japonica*. There is also a form with strikingly variegated leaves, which is worthy of cultivation as a foliage plant.

Iris Fimbriata

Iris albicans Lange
figured as *Iris florentina*

It is unfortunate that the plant illustrated so beautifully by Redouté was not the same plant as the Florentine Iris, from the rhizome of which the perfume and medicinal agent known as orrisroot was obtained. The plate actually depicts an Arabian species, which was later described by J.M.C. Lange as *Iris albicans*. The true Florentine Iris is the more well known of the two and is also white-flowered, but is almost certainly nothing more than a colour form of the common *Iris germanica* and not a distinct species.

W.R. Dykes, in his superb monograph, *The genus Iris*, sets out the differences between the two plants and there can be no doubt about the identity of Redouté's plant. The pure white flowers immediately attract attention since those of the Florentine Iris have a very slightly pale blue tint. The greyness of the rather broad leaves, noted in the original text for this plate as being rather unlike those of the 'Iris de Germanie', give an additional clue. It is, however, the way in which the flowering stem branches that provides the more definite proof of identity. The lowest flower is shown to be, and is described as, sessile, whereas in *Iris germanica* and its white variety the lower branch is some 8 to 10 centimetres long.

In Arabia and Yemen where this plant is apparently a native it occurs in both white and blue forms. It is fairly certain that the white form was distributed throughout the Muslim world, as a plant with which to adorn cemeteries, white being the colour of mourning, and it has long survived in Spain and Greece the disappearance of Islam.

Iris Florentina

IRIDACEAE

Iris foetidissima Linnaeus

GLADWYN

The Gladwyn, Stinking Iris or Stinking Gladdon has tough evergreen leaves which release an unpleasant odour when bruised, but in spite of this undesirable feature it is a most useful and attractive garden plant. The flowers are rather dull in their colours, although there is a yellow-flowered variant—variety *citrina*—whose blooms are much more desirable than the usual muddy purplish form. However, it is in the fruiting stage in mid-winter that *Iris foetidissima* really comes into its own; its large capsules split open to reveal brilliant orange-scarlet seeds which remain attached for a long period, even when cut and dried. Also, the evergreen leaves make quite striking clumps and there is a striped variegated form which is of value in dull shady corners of the garden.

This curious iris has no near relatives in the genus and is placed by most authorities in a section of its own. It is a native of western Europe from Spain northwards to England, where it occurs in woods particularly on the chalk and limestone hills.

Another name commonly used for this plant is Roast Beef Plant. The smell of the leaves has been likened to roast beef, boiled milk and wet starch, three quite dissimilar aromas one might think. The name Gladwyn is probably derived from the Latin *gladius*, sword, referring to the sword-shaped leaves.

1. Outer perianth segment 2. Ovary and style branches
3. Capsule 4. Single seed

Iris Foetidissima

Iris germanica Linnaeus

GERMAN FLAG IRIS

The origin of *Iris germanica* is a mystery which will probably never be solved. It is an extremely widespread plant of some antiquity, certainly dating back several centuries, and probably one of the oldest and most tenacious of a group of bearded iris hybrids. It is naturalized in many countries and is vegetatively very robust, so that it can persist and indeed increase without the necessity for seed production. W.R. Dykes reports that in the warm climates of the Mediterranean region it does, however, produce capsules and that the offspring vary somewhat in depth of colour. In the cooler climates of northern Europe and North America seeds are rarely produced.

The most common form of *Iris germanica* seen today is of a blue-purple colour, the standards paler than the falls. The bracts are partly green and partly papery and are often tinged with purple, quite unlike the silvery ones of *Iris pallida* for example. Linnaeus gave the name *Iris germanica* to an iris of similar appearance, but whether or not it was the same clone as our present-day plant it is impossible to say. *Iris germanica* may of course be truly native in some part of the Mediterranean, but its distribution is now so wide—almost worldwide in fact—that any speculations must be of the most tenuous.

Iris Germanica

IRIDACEAE

Iris graminea Linnaeus

One of the most notable features of this fairly modest iris is the very strong fruity fragrance; this more than compensates for the small size and subdued colour of the flowers. It has been likened to the scent of fresh bananas, plums, and greengages. In fact one of its Latin names in the seventeenth century was *Iris angustifolia prunum redolens*, the plum-scented narrow-leaved iris.

Iris graminea has been known to European horticulturists for centuries since it is a widespread native of central and southern Europe and is extremely easy to grow in almost any garden situation. The normal flowering time is May or June.

It belongs to the large group of irises named by J.G. Baker the Apogon section or beardless irises because, unlike the bearded Pogon section, these have no hairs on the falls. The only European representative of this group to possess a fruity scent, *Iris graminea* is also distinguished by having a very strongly flattened, almost winged, stem. The related and similar *Iris sintenisii* from the Balkans has a much more cylindrical stem and the violet-blue flowers are without a scent.

The Apogon section can be further subdivided into several groups, and *Iris graminea* probably belongs best with the Spuria, or Salt-Marsh Irises, although these do not have flattened stems or a fruit-like scent, and, unlike the majority of them, *Iris graminea* is not an inhabitant of open salt-marsh or sand-dune country.

1. Outer perianth segment with stamen
2. Ovary and style branches

Iris Graminea

Iris spuria Linnaeus subspecies *ochroleuca* (Linnaeus) Dykes

figured as *Iris ochroleuca*

SALT-MARSH IRIS

If, as is the case in the recently published *Flora Europaea* (Volume 5, 1980), *Iris ochroleuca* is treated as a subspecies of the extremely widespread *Iris spuria*, then the familiar epithet can remain unchanged under the International Rules of Nomenclature. However, if the plant is regarded as a separate distinct species, an unfortunate change to Philip Miller's older name of *Iris orientalis* is required by the Rules. The change would not matter unduly were it not for the fact that there is another familiar but quite different species long known in horticulture as *Iris orientalis.*

The taxonomy of the Salt-Marsh Iris and its relatives is poorly understood and is in need of a thorough investigation. Until such time as this study can be undertaken, it seems scarcely necessary to change such a well-established name as *Iris ochroleuca.* Suffice it to say that it is a striking plant, well worth cultivating and it thrives in almost any sunny position. It is often seen growing in great quantities along the irrigation ditches and riversides in Turkish Anatolia, where the soils are usually rich in salts. However, a lack of these in garden soils does not seem to discourage its growth.

Iris ochroleuca is one of the taller members of the Spuria group and in its white and yellow coloration can be easily distinguished from the rest, which usually have either silvery lavender, violet or deep-yellow flowers.

1. Outer perianth segment and stamens 2. Ovary, style branches 3. Capsule

Iris Ochroleuca

Iris pallida Lamarck

The tall bearded irises are undoubtedly the most popular plants in the whole genus for general garden use, and the many cultivars now available far surpass the wild species in their range of flower colour and size. The origins of many of these hybrids are obscure and it is impossible even to guess at the parentage, but of one fact we can be sure: *Iris pallida* played a great part in the early development of these lovely plants.

In distinguishing between the various species in the bearded, or Pogon, group of irises, the colour and texture of the bracts which enclose the flower buds is rather important. *Iris pallida* has bracts, or spathes as they are sometimes called, which are wholly papery and silvery white. Thus, they are quite unlike all the other species, in which the bracts are at least partly green or tinged with purple.

Iris pallida is a vigorous species with a branched stem reaching 1 to 1.5 metres tall, and with a fan of robust greyish leaves. A well-flowered clump is an impressive sight and equal in beauty to any of the garden varieties. The soft lilac-blue flowers have the additional bonus of being strongly but delicately scented.

In the time of Redouté it was assumed that *Iris pallida* came from 'the Levant', but it is now known that it occurs on the mountains along the Adriatic coast of Yugoslavia and northern Italy, where it sometimes grows in large numbers amid the limestone rocks and scrub.

1. Outer perianth segment
2. Ovary, style branches and stamen

Iris Pallida

Iris pseudacorus Linnaeus

YELLOW WATER FLAG

This common waterside iris is a familiar plant throughout Europe and occurs also in North Africa, Turkey, Iran and Russia, often lining the banks of streams, ditches and canals with its bright yellow flowers in summer. It occurs too in marshy places near lakes and ponds and can cover extensive areas. The name 'false acorus' refers to the similarity between its foliage and that of Sweet Flag, *Acorus calamus*.

The Yellow Flag or Yellow Iris belongs to the beardless group of *Iris*, or Apogon section as it is known, which encompasses those irises that do not have a tuft of hairs on the falls. Within the Apogon section there are some natural subdivisions which enable the many species to be grouped together. *Iris pseudacorus* is apparently most closely related to the eastern Asiatic *Iris laevigata* and *Iris kaempferi*; these are also moisture-loving plants, but are immediately separable from the species illustrated here by their non-yellow flowers. *Iris versicolor*, with its variants which are sometimes described as separate species, is the final member of this small group. A native of the eastern United States, it has bluish or purplish, strongly veined flowers and grows, too, in just the same type of habitat as *Iris pseudacorus*.

The Yellow Flag has been used as a purgative, apparently very effectively, for William Meyrick noted that it 'has been found to procure plentiful evacuation from the bowels when all other means have proved ineffectual'.

Iris pseudacorus varies a little in its coloration and some of the variants have been selected for horticultural purposes; for example, the pale yellow form without the darker blotch on the falls, which has been given the name variety *bastardii*. The amount of brown or violet veining also varies somewhat.

Ovary, style branches and stamen

Iris Pseudacorus

Iris sambucina Linnaeus

When Linnaeus first described this bearded iris in 1759, he gave a very scanty description, quite insufficient for us to identify the plant he had in mind. However, in 1762 he expanded this considerably in the second edition of the *Species Plantarum*, and described *Iris sambucina* as having violet or bluish flowers with bluish style branches. The plant closely resembles *Iris squalens*, also illustrated by Redouté (page 185), but has very little of the dingy yellowish coloration which is such an obvious feature of *Iris squalens*.

Almost certainly *Iris sambucina* and *Iris squalens* are hybrids, both of which come from the same parent species. However, the form of *Iris variegata* involved in the production of *Iris sambucina* was one without a strong background of yellow. The Variegated Iris, figured on page 189, is an eastern European species and varies considerably in the wild, from forms with a very yellowish appearance to ones in which the ground colour is nearly white. It is believed that the other parent of *Iris sambucina* is *Iris pallida*. If this is true, it accounts for the rather bluish appearance of the flower.

W.R. Dykes, in his *Handbook of Garden Irises* (1924), offers some fairly convincing evidence concerning the parentage of both *Iris sambucina* and *Iris squalens*. He states that above Carlopago (now Carlobag) on the Adriatic coast, he has found both *Iris pallida* and *Iris variegata*, and also hybrids between the two. 'The same two species occur together near Bozen [now Bolzano] with similar hybrids, which were once known as sambucina and squalens.'

1. *Outer perianth segment*
2. *Ovary, style branches and stamen*

Iris Sambucina

IRIDACEAE

Iris sibirica Linnaeus

SIBERIAN IRIS

This popular garden plant is a widespread native of central and eastern Europe from France to Poland and Russia, extending southwards into the Caucasus and northeast Turkey. It usually occurs in damp grassland or in rich streamside vegetation, but in cultivation seems to be perfectly at home in almost any situation, given sufficient moisture in the growing season.

Iris sibirica belongs to a group of closely related beardless species within the genus, known collectively as the series Sibiricae. They are characterized by being tall slender plants with hollow stems (except in *Iris clarkei*) and by having a seed capsule which is three-cornered in cross-section. The shape of the stigma, rather a useful feature in grouping the beardless iris species, is like a triangular tongue. The ten species are, apart from *Iris sibirica* which extends westwards into Europe, entirely Asiatic and mostly from China and the Himalayas.

The nearest relative to the confusedly named Siberian Iris (it does not actually occur in Siberia!) is *Iris sanguinea* (*Iris orientalis* of C.P. Thunberg), which occurs in eastern Russia, China and Japan. The main difference lies in the bracts, which subtend the flowers, those of *Iris sibirica* being brown and membranous at flowering time and those of *Iris sanguinea* being green when the flowers are fully expanded. Also, the stems are usually unbranched in *Iris sanguinea* and carry only two flowers at the apex, whereas they are normally branched in *Iris sibirica*, with three to five flowers.

In the wild the Siberian Iris is normally violet-blue with occasional albinos, and it is one of the latter forms that Redouté has illustrated as *Iris sibirica* variety *B*. As a result of much selection, the species does however vary considerably in the garden forms, in both colour and size of flower. Nearly all the forms show the beautiful netted pattern of veins on the lower part of the falls which is one of the main attractions of the species. *Iris sibirica* also crosses with other members of the group and there is a wide range of colourful hybrids now in cultivation.

1. Ovary, style branches and stamen 2. Capsule

Iris Sibirica

IRIDACEAE

Iris sibirica Linnaeus
figured as *Iris pratensis*

SIBERIAN IRIS

Iris Pratensis

Iris planifolia (Miller) Fiori & Paoletti
figured as *Iris scorpioides*

This beautiful little winter-flowering iris has several very distinctive features which make it instantly recognizable in a large and diverse genus. The papery-coated bulb, thick fleshy roots, channelled leaves and reduced inner tepals, or standards, immediately link it with the interesting Juno group of species, which are mainly distributed in Iran, Afghanistan and central Asiatic Russia.

The only near relative to *Iris planifolia* in the Juno group is the Palestinian *Iris palaestina*, which is rather like a greenish version of it. These two species are distinctive in having pollen grains covered with minute spines, thus distinguishing them from the more easterly-occurring species which have smooth pollen. Within the Juno group these two have been placed in the aptly named Acanthospora section.

Iris planifolia is the only species of the group to occur in Europe and North Africa and is not uncommon in parts of southern Spain, Sicily and Algeria. It also occurs in Portugal, Sardinia, Crete, Libya and Tunisia, where it is on the whole less frequently seen. In the main areas of its distribution it flowers in November and December at low altitudes on rocky slopes, but it may appear as late as February in some districts.

It has a long history in cultivation and was known to Carolus Clusius in 1576 as *Iris bulbosa latifolia*, a far more useful descriptive name than *Iris planifolia*, meaning flat-leaved, which it is not! J.L.M. Poiret's name, *Iris alata*, winged iris, is equally obscure, but possibly refers to the wide wings on the outer tepals, or falls. Another synonym, *Iris scorpioides*, is as unhelpful as it is unpleasant for there is little that is obviously scorpion-like about the plant.

The flowers of *Iris planifolia* are normally of a mid-lilac-blue, but there are forms of a darker violet-blue and occasionally pure white ones. Another feature of this charming member of a superb genus is the strong perfume, an extra bonus in the depth of winter.

1. Style branches with stamen
2. Inner perianth segment

1 2

Iris Scorpioides

IRIDACEAE

Iris spuria Linnaeus

Under the broad umbrella of the name *Iris spuria* comes a wide range of irises whose true status is to this day a puzzle. From Denmark and England, south to the Mediterranean and eastwards into Afghanistan and China these Salt-Marsh Irises occur, often along the edges of irrigation ditches or in grassy plains which do not dry out excessively in the summer months. In the more northerly and colder climates, *Iris spuria* is however associated with maritime dunes. There are many specific names attached to the numerous variants and it is probable that there are indeed several distinct species within the group loosely known as the Spurias. Work by Dr L.W. Lenz has shown that there are several different chromosome numbers in the group and these are associated with the geographical distribution, which also suggests that *Iris spuria* should not be treated as one widespread and variable species.

In the strict sense, *Iris spuria*, as described by Linnaeus, is the central and northern European plant depicted by Carolus Clusius as long ago as 1601 in his *Rariorum Plantarum Historia*. Linnaeus apparently believed that it was a hybrid between *Iris foetidissima* and *Iris graminea* and accordingly gave it the epithet *spuria*.

The wild *Iris spuria* of Europe is usually lilac or violet-blue with a little variation in depth of colour, the form illustrated here being one of the slightly darker ones. In Asia, however, there is much variation, from white through all shades of silvery lavender to deep violet-blue and yellow. These variations account for at least some of the different names associated with Spuria irises. The ability to vary and to hybridize has been used to advantage by plant breeders, particularly in America where a wide range of attractive cultivars now exist, ranging through many shades of blue, lilac, violet, yellow, orange, bronze and white.

1. Ovary, style branches and stamen 2. Capsule

Iris Spuria

Iris squalens Linnaeus

This is yet another of the tall bearded irises, the origins of which are lost in antiquity and were probably never on record anyway. Irises of this group have been cultivated for many centuries and the multitude of hybrids now in existence are derived from only a few wild species. It is unlikely that plants such as *Iris squalens* and the similar *Iris sambucina* were the result of controlled crosses between wild species; more probably they were chance hybrids which arose either in the wild or in the collections of iris enthusiasts.

Both of the plants mentioned above were described by Linnaeus in 1759, but they were probably known in gardens well before this. The description given by Linnaeus was at first very vague but he later amplified it, so we can be quite sure that the plant illustrated here by Redouté is in fact *Iris squalens*.

Although we can only make well-informed guesses as to the parentage of plants such as this, it seems very likely that W.R. Dykes is correct in his supposition that *Iris squalens* was derived from *Iris pallida* and *Iris variegata*, which are illustrated on pages 179 and 189. The veining on the falls is very reminiscent of the latter and the dingy nature of the yellowish standards could well be caused by the merging of the pale lavender-blue of *Iris pallida* with the yellow of *Iris variegata*. *Iris squalens* does not in fact show any characteristics which detract from this suggested parentage.

1. Outer perianth segment
2. Ovary, style branches and stamens

Iris Squalens

Iris susiana Linnaeus
figured as *Iris suziana*
MOURNING IRIS

This extraordinarily coloured iris belongs to the group of species known as the Oncocyclus section, which is characterized by having rhizomes, solitary flowers with a beard on the outer three segments, or falls, and seeds which have a fleshy white attachment known as an aril. All the species constituting this rather exotic-looking group are distributed in the Near East, from central Turkey and the Caucasus, southwards to Israel and eastwards to northeast Iran. The species that occur in the more southerly regions, that is those from Syria, Lebanon and Israel, are on the whole more robust and larger-flowered plants than their more northerly relatives.

With its huge speckled and veined flowers and wide leaves, *Iris susiana*, or the Mourning Iris as it has become known, undoubtedly belongs to the southern group. Exactly where this plant originated is not known, however, for it reached Europe in 1573 by way of Istanbul (then Constantinople), having been sent from there by Busbecq. Nothing similar exists in Turkey today and it can safely be assumed that it originally came from farther south and was probably cultivated by the very garden-conscious Ottoman Turks of that period. To this day the name for iris in Turkey is 'süsen'.

The most likely explanation of the origin of *Iris susiana* is that it is a selection of one of the several very variable wild species such as *Iris sofarana* or *Iris basaltica*, which occur in Syria and Lebanon. Like most other members of the Oncocyclus section, *Iris susiana* needs the protection of a frame to survive and flower in northern gardens, but it is a very successful and vigorous plant in countries with warm dry summers.

The sectional name Oncocyclus (from Greek *onkos*, mass, swelling tumour, *kuklos*, circle) refers to the aril on the seeds.

Iris Suziana

Iris variegata Linnaeus

VARIEGATED IRIS

This rather striking bearded iris has a long history in cultivation for it was known to Matthias de Lobel and Carolus Clusius in the sixteenth century. It has undoubtedly given rise to some of the exotic-looking tall bearded garden irises that we know today and was probably one of the parents of *Iris sambucina* (page 175) and *Iris squalens* (page 185).

Variegated Iris is native in eastern Europe and it varies considerably over its wide area of distribution. The plant illustrated here, with a yellow ground colour and purplish veining, is just one of the forms which occur and it is possible to find even more strikingly coloured ones, in which the veins run completely together on the blade of the falls into a rich reddish-purple stain. Other, rather more rare, forms have a whitish background to the veining and a Yugoslavian collection of such a plant was at one time separated as a distinct species, *Iris reginae*. However, there seems little justification for giving these different colour forms separate specific names, important as they are from a gardener's point of view.

Within the bearded group of irises, *Iris variegata* is distinct and easily recognized. Apart from its flower colour it is the only species in which the stem is rather short, usually about 30 to 45 centimetres, but at the same time branched. The leaves are strongly ribbed and green rather than the greyish colour of many bearded iris leaves.

Iris Variegata

IRIDACEAE

Iris versicolor Linnaeus

This very widespread North American iris was so named because of the mixture of colours in the individual flowers, although it is also apt on account of the very variable overall flower colour that exists over its vast natural territory. *Iris versicolor* inhabits streamsides and swampy places from eastern Canada southwards to Texas, and over this huge area the blade of the falls can be blue, purple, reddish or dull slate coloured. A really good reddish-purple form, variety *kermesina* is often cultivated.

Like *Iris pseudacorus*, the common Yellow Flag Iris of European watersides, this American species is an ideal plant for cultivating on the margins of ornamental streams or ponds. Indeed, it is so easy to grow that in a few countries it has actually left the confines of the garden and become naturalized. Although it is undoubtedly at its best near water, it can also be grown as a herbaceous border plant, given plenty of humus in the soil. In the plant described as *Iris virginica*, or *Iris versicolor* variety *virginica*, the principal colour is blue-purple rather than the claret or reddish colour of the more commonly encountered form, but the various forms are all equally ornamental. Typically, the species is about 60 centimetres or more high when in flower, but dwarf forms with blue flowers are known.

The rhizomes and roots of *Iris versicolor* have been used for medicinal purposes in the past, an extract from them having been found beneficial in the treatment of diseases of the liver and in dropsy.

Iris Versicolor

IRIDACEAE

Iris latifolia (Miller) Voss
figured as *Iris xyphioides*

ENGLISH IRIS

The English Iris belongs to the relatively small group of bulbous irises known as the Xiphium section; these are confined to the western Mediterranean region and are predominantly Spanish and Portuguese. Most of the seven species occur in hot sunny places or at least in habitats that dry out in the summer months. *Iris latifolia*, on the other hand, is a plant of damp grassy places in northern Spain and the Pyrenees and is consequently more tolerant to cultivation in the gardens of northern Europe.

It has been cultivated for several centuries and there are now many attractive garden selections in shades of blue, violet or purple; albino forms are also popular. Unlike its near relative *Iris xiphium*, the Spanish Iris, it never produces the yellow forms which are so frequently seen in that species. There is a marked difference in flower shape between the two and although the names are similar there can be no mistaking the large-flowered English Iris. The lower part, or haft, of the three outer petals, or falls, is widened so that it is wing-like, whereas in *Iris xiphium* the margins of the haft are more or less parallel and not enlarged. Moreover, the leaves of *Iris latifolia* do not appear until the spring, whereas in the Spanish Iris they are already well developed by the winter. The tube of the flower in both of these species is only 6 millimetres long or less, which distinguishes it from all the other species in the Xiphium section.

The English Iris owes its name to the fact that Carolus Clusius, in the late sixteenth century, was sent some bulbs which were being cultivated near Bristol. This was an important trading centre for Spanish goods and no doubt the bulbs originated from the Pyrenees but the name English Iris has been attached to *Iris latifolia* from that day to this.

Bulb

Iris Xyphioides

IRIDACEAE

Chasmanthe aethiopica (Linnaeus) N.E. Brown
figured as *Antholyza aethiopica*

The specific epithet *aethiopica* may today mislead one into assuming that this plant is a native of northeast tropical Africa, but it dates from a time when 'Aethiopia' usually referred to the whole of Africa south of Egypt. *Chasmanthe* is in fact an entirely South African genus, consisting of about ten species. *Chasmanthe aethiopica* occurs at fairly low altitudes, often on sandy flats and dunes in hedgerows and woods in the winter-rainfall region of the Cape, where it flowers in June and July.

The chasmanthes are characterized by having very irregular-shaped flowers with a long, often rather inflated tube above a narrow base, and a large hooded upper perianth segment.

The genus is closely related to *Crocosmia*, which is commonly cultivated as Montbretia in European gardens, and to *Tritonia*, another South African group of plants. Dr P. Goldblatt, who has made a thorough study of the genera of Iridaceae of southern Africa regards these as forming a natural group, which he calls the tribe Tritoniineae. He considers *Chasmanthe* to be the most highly modified of the genera in this group and certainly their flowers are of a very complex appearance.

Unlike the Montbretia, *Chasmanthe aethiopica* is not hardy in most northern European gardens although it can be seen in great numbers in certain areas of southwest Britain and in the Mediterranean region.

1. Flower 2–3. Bracts 4. Pistil 5. Corm

Antholyza Aethiopica

Belamcanda chinensis (Linnaeus) de Candolle

BLACKBERRY LILY

This showy member of the iris family is very closely allied to the true irises and indeed will hybridize with *Iris dichotoma* to produce a remarkable range of intermediate offspring. *Belamcanda chinensis* is the only species in the genus. The flower, with its six equal perianth segments and three slender style branches, is in fact quite different in structure from that of an iris, in which the six perianth segment are separated into falls and standards and the style branches are flattened and petal-like. The fruits also are rather distinctive, with capsules opening to reveal large black seeds; hence the name Blackberry Lily used in some countries.

Like *Iris, Belamcanda* produces a fan of flat leaves from a small rhizome and in summer sends up a branching flower stem from the centre of the leaf cluster. This stem can reach 2.5 metres in wild specimens, but it is usually much less than this in cultivation. Each flower is of rather short duration, but because there is a succession of them, quite a striking display is produced over a considerable period of time. Although individual plants are usually short-lived, seeds are freely produced and the young plants rapidly reach maturity.

Belamcanda is a native of China, Japan, Taiwan and the Himalayan region. The root has been used to cure sore throats and fevers and is also recommended as an antidote to poisons, in particular the bite of a cobra.

1. Fruit 2. Axis of fruit 3–4. Seeds

Belamcanda Chinensis

IRIDACEAE

Crocus flavus Weston
figured as *Crocus luteus*

YELLOW CROCUS

To discover the original cultivation of this beautiful spring crocus we must, as with many other bulbous plants from Europe and Asia Minor, turn to Carolus Clusius. Over four hundred years ago, in March 1579, he received from Serbia the plant which he named *Crocus vernus latifolius flavoflore* and it is reasonable to assume that its offspring were subsequently distributed through Europe. John Gerard in his *Herbal* of 1597 noted that some of the crocuses he knew had been sent by Clusius and that one of them had 'flowers of a most perfect shining yellow colour, seeming a far off to be a hot glowing cole of fire'. He had received it via Jean Robin of Paris and it is exciting to think that from a single introduction by Clusius, the plant may have been passed from one great gardener to another to become one of the most loved of all spring bulbs.

Over the years, horticulturists selected and hybridized *Crocus flavus* and at one time there were many forms, or cultivars, differing in depth of colour from creamy white to deep orange, some with dark striping on the exterior. Most of these are now lost to cultivation, but none ever surpassed the quality of the brilliant orange-yellow wild form. The most commonly grown variant, which is now widely available from nurserymen as 'Dutch Yellow', 'Golden Yellow', 'Large Yellow' or 'Yellow Giant', is of hybrid origin, the Crimean *Crocus angustifolius* (*Crocus susianus*) being its other parent. This vigorous hybrid is a most satisfactory garden plant since, although sterile, it increases very rapidly by corm division and is hardy to the point of being indestructible.

In the wild, *Crocus flavus* inhabits the Balkans and western Turkey, preferring light woodland or scrub at varying altitudes from sea level up to 1,000 metres. It flowers in March and April.

1. Flowers and bracts 2. Flower opened out to show stamens 3. Pistil

Crocus Luteus

IRIDACEAE

Crocus sativus Linnaeus

SAFFRON CROCUS

This beautiful autumn-flowering crocus is the source of saffron, long recognized as a brilliant yellow dye and flavouring agent and much used in the past for its alleged medicinal properties. It was the first plant to be named 'Crocus' (Greek *krokos*), a name of great antiquity.

The three long, deep-red style branches contain the saffron and herein lies the reason for its costliness; it is said that as many as four thousand three hundred flowers are required to produce 28 grams (1 ounce) of the dried styles. Because it is not a good economic proposition in our day, the cultivation of *Crocus sativus* for saffron has died out in most countries. However, it still goes on in a few places—Safranbolu in Turkey, for example, where the climate is ideal for free flower production. In many European countries, where it was formerly grown commercially, the cultivation is not a problem, but the plants bear few flowers. Undoubtedly the climatic changes since the Middle Ages, and cooler summer temperatures, have caused this loss of flowers, certainly in countries such as England, where saffron production petered out in the eighteenth century. *Crocus sativus* is unknown as a wild plant, although it closely resembles the smaller Greek species *Crocus cartwrightianus* and is probably a variant of this, selected for its vigour and improved size of style.

Saffron has been used as a dye since early Greek times, probably as early as 1600 BC, for at Knossos in Crete there exist designs on Minoan frescoes and pottery showing a crocus similar to the plant grown today.

In ancient times it was apparently strewn on the floors of halls, courts and theatres as a perfume. Moreover it played such an important part in herbal medicine that in the seventeenth century Johann Ferdinand Hertodt wrote a book, *Crocologia*—almost three hundred pages of instructions on the uses of saffron in the treatment of diseases. It was recommended for almost any complaint and Hertodt claimed that it could cure such diverse disorders as hernia and cerebral haemhorrage.

Flower opened out to show stamens and pistil

Crocus Sativus

IRIDACEAE

Crocus vernus (Linnaeus) Hill

SPRING CROCUS

Much has been written, and doubtless a great deal more will follow, on the naming of the common spring crocuses of the Alps and Italy. The simple answer, perhaps over-simplified, but practical, appears to be that there are two entities involved: the tiny white crocus which has its stigma overtopped by the stamens, and the larger purple one in which the style is prominently projecting above the stamens. Unfortunately the former sometimes produces small purple or striped forms and the latter can occur in albino forms. Such are the vagaries of plants!

Although the two overlap in their distribution, it can be said that the small white crocus often occurs in exposed mountain turf, generally at higher altitudes than the larger one, which is frequently to be found in sparse woodland. Where they meet the occasional hybrid can be found with curiously striped flowers.

The search for the correct names for these two plants has occupied the time of many a botanist, and no doubt the arguments will continue unabated. However, I can see no reason for rejecting the findings of B.L. Burtt who, in E.A. Bowles's *Handbook of Crocus and Colchicum* (1952), argued the case for accepting the name *Crocus vernus* of John Hill for the common alpine crocuses. In the recently published *Flora Europaea* (Volume 5, 1980) I have divided this into two subspecies: *vernus* for the larger purple crocus and *albiflorus* for the small white one.

Redouté's illustration shows the purple and white forms of the larger *Crocus vernus* subspecies *vernus.*

1–2. Colour variations 3. Pistil
4. Flower opened out

Crocus Vernus

IRIDACEAE

Ferraria crispa Burman
figured as *Ferraria undulata*

SPIDER FLOWER

This small genus of strangely coloured iris-like plants is named after G.B. Ferrari who, in the middle of the seventeenth century, was the first to publish a description and illustration of one of the species. It was in fact the species shown here, although he named it *Flos indicus e violaceo fuscus*. Redouté in his original plate followed Linnaeus in calling the plant *Ferraria undulata* but J. Burman's name *Ferraria crispa* is the first validly published binomial after 1753, the earliest date from which names are acceptable under the International Rules of Nomenclature.

Ferraria crispa is probably the best known of the ten species in the genus; it was introduced into Europe at a very early date and has probably been in cultivation for about three hundred years. All ten occur in southern Africa, most of them in the winter-rainfall region of the southwest Cape, and are winter- or spring-flowering both there and when cultivated in the northern hemisphere. The species illustrated here occurs in rather sandy lowland areas of the Cape, usually in coastal regions. It is therefore not very hardy in most gardens in Europe and Britain and has always been treated as a cool greenhouse plant.

Ferrarias have a very curious underground structure consisting of a string of flattish corms, a new one being added each year. The stems are entirely sheathed by the leaves which are carried in one plane, that is, distichously arranged. These features and the strikingly fimbriate branches of the style make *Ferraria* easy to distinguish from all the other related genera such as *Moraea*, *Homeria* and *Dietes*.

The flower colour of *Ferraria crispa*, or Spider Flower as it is sometimes known locally, varies greatly, from dark reddish-brown to blackish to creamy yellow with dark stripes and blotches. With this colouring it is not surprising that it also has an unpleasant, almost putrid smell and is pollinated by flies.

1. Pistil and stamens   2. Pistil

Ferraria Undulata

IRIDACEAE

Gladiolus cardinalis Curtis

WATERFALL GLADIOLUS

This colourful South African species bears the picturesque local names of Waterfall Gladiolus or Nuweyaarsblom, the New Year Lily. In its native habitat it is a plant of damp rock ledges and it flowers in mid-summer, that is December or January. Of course, when introduced into gardens of the northern hemisphere it adjusts its cycle of growth by about six months, to flower in July or August.

No doubt because of its spectacular appearance, *Gladiolus cardinalis* was one of the first species to be brought into Europe, and it thus also became one of the parents of the earliest known hybrids. As early as 1790 the species was illustrated and described in the *Botanical Magazine* from specimens flowering in an English nursery which had obtained the corms from Holland. Although itself rarely grown in gardens today, its offspring are well known and include the lovely *Gladiolus colvillei* hybrids.

It is unfortunate that the many spectacular florists' gladiolus of today have quite overshadowed the wild species in gardens and it is almost impossible to obtain this or any other of the one hundred or more South African species through commercial sources.

Gladiolus Cardinalis

IRIDACEAE

Gladiolus communis Linnaeus

COMMON WILD GLADIOLUS

The temperate northern hemisphere species of *Gladiolus* are not very numerous compared to those of tropical and South Africa but they are rather tricky to identify. The species are individually very variable and often appear to be linked by intermediate forms; also, they will hybridize wherever their distributions overlap. *Gladiolus communis* is very closely allied to *Gladiolus illyricus* and to the much-cultivated *Gladiolus byzantinus.* Although there is an overlap in the characters, it is generally possible to separate the Illyrian Gladiolus from the other two by its overall stature, leaf dimensions and rather few-flowered unbranched inflorescence. It seldom exceeds 50 centimetres in height and may have as few as three flowers, whereas *Gladiolus byzantinus* and *Gladiolus communis* frequently exceed 60 centimetres and often have branched flower stems with more than ten flowers on each branch. Unfortunately Redouté chose to illustrate an unbranched specimen of the latter.

Between *Gladiolus communis* and *Gladiolus byzantinus* there is an even less clear distinction and A.P. Hamilton regards the Byzantine Gladiolus as a subspecies of the former rather than a distinct species (*Flora Europaea*, Volume 5, 1980). *Gladiolus communis* is generally a little less robust and has flowers which tend to be pinkish-red rather than the deep reddish-purple of *Gladiolus byzantinus*, but it seems there is no individual feature that can be depended on to distinguish between them.

Gladiolus communis is a plant of dryish field and scrub in Mediterranean Europe from Spain to Greece, and probably Asia Minor also, whereas *Gladiolus byzantinus* in spite of its name is mainly a North African plant although present in Spain and Sicily too.

Gladiolus Communis

Gladiolus undulatus Linnaeus
figured as *Gladiolus cuspidatus*

This is a widespread but rather rare gladiolus from South Africa, where it occurs mainly in the southwestern Cape area. It varies considerably from place to place and the form figured by Redouté is only one of several rather different-looking plants. In the excellent *Revision of the South African Species of Gladiolus* of 1972, G.J. Lewis and A.A. Obermeyer record that it can be white, cream, greenish-cream or pinkish, with narrow or broad leaves and varies in stature from about 30 centimetres to well over a metre tall. The markings on the flowers can be rather faint or very prominent and the lower perianth segments can have either a purplish line along their centres or a dark spear-shaped mark. The segments are often long-tapering to the apex, but sometimes they are less acutely pointed.

Gladiolus undulatus has been known to botanists for a very long time, the first record in Europe being in 1764 when Linnaeus received some dried specimens collected in South Africa by the Burman expedition. It has been cultivated more widely than most species but is rarely seen today in northern hemisphere gardens in spite of being hardy as far north as southern Britain. The species has become naturalized in several countries, for example in Australia and Mauritius, and in favourable conditions can be a weed, with its corms producing a large cluster of young cormlets each year.

1. Perianth opened out 2. Pistil 3. Corm

Gladiolus Cuspidatus

Freesia refracta (Jacquin) Ecklon ex Klatt
figured as *Gladiolus refractus*

The well-known genus *Freesia* is, in spite of its spelling, named after Dr Friedrich Heinrich Theodor Freese, a physician at Kiel. He was a friend of C.F. Ecklon, who had obviously intended to name the plant after him, in which case it would have been spelled '*Freesea*'. Unfortunately he never published the name validly and F.W. Klatt eventually described the genus in 1866 using the spelling we know today.

Freesia refracta was not the first species to be cultivated in Europe. This distinction goes to *Freesia caryophyllacea*, which was recorded as being in cultivation in 1759, from bulbs sent to Johannes Burman. Nearly thirty years later, in 1786, the species *Freesia refracta* was described, under the name *Gladiolus refractus*, by Nicolaus Jacquin.

The easy cultivation, free-flowering habit, showy flowers and delicate fragrance made these two species and other freesias popular garden plants, especially in the late nineteenth and early twentieth centuries when many hybrids were created. Although the genus is quite a small one, containing perhaps only twenty species, the range of size and colour through hybridization and selection is now enormous.

Freesia refracta, one of the parent species of our florists' freesias, is in its wild form not a showy plant. It has smallish but very fragrant greenish-yellow to pale yellow flowers, suffused with dull purplish markings on the lower perianth lobes. In the southwest Cape region of South Africa it flowers between July and October and inhabits stony or rocky hillsides, usually at fairly low altitudes and occasionally in dunes at sea level. The illustration by Redouté is not quite typical in that the flower spike normally bends abruptly at its base, allowing the flowers to face upwards.

1. Corm 2. Capsule and seed

Gladiolus Refractus

Gladiolus carinatus Aiton
figured as *Gladiolus ringens*

This slender South African species of *Gladiolus* is notable for its delicate scent, a feature that has been used to advantage by Professor T.T. Barnard of Dorset in England who has produced some attractively scented gladiolus hybrids.

The Sandpypie or Blue Afrikaner is a very common species in the southwestern Cape area, where it grows in sandy places and on mountain slopes, flowering between July and September. Thus, it is naturally a winter-flowering plant and in cultivation in the northern hemisphere continues this habit and flowers in late winter, usually in February or March. In all but mild areas it and its hybrids therefore require frost-free conditions, but they are charming plants and easy to grow in cool greenhouses or conservatories.

In the wild, *Gladiolus carinatus* varies greatly in colour, from pale blue to deep violet, purple, pink or even a reddish shade, and there are yellow or brownish forms too. The lower lobes of the perianth have yellow bands along the middle.

One striking feature of the plant is the spotting on the sheathing leaves as they emerge at ground level, and this characteristic prompted William Aiton, when he published the species in 1789, to invent the common name Spotted-Stalked Corn-Flag. The specimen mentioned was cultivated at Kew Gardens at the time, the corms having been sent by Francis Masson some years earlier. Thus it is a plant with a long history of cultivation, although its tender nature has ensured that it has always been one for the specialist gardener.

1–2. Bracts 3. Flower
4. Stem 5. Perianth 6. Pistil

Gladiolus Ringens

<u>*Babiana tubulosa*</u> – (Burman fil.) Ker-Gawler – variety <u>*tubiflora*</u> (Linnaeus fil.) Lewis

figured as *Gladiolus tubiflorus*

The iris family which contains many diverse-looking plants such as *Gladiolus*, *Tigridia* and *Crocus* also includes the large group of plants known as *Babiana*. When in flower, these have a definite similarity to the more well known freesias, although the tough pleated leaves, often hairy on the surface, give the babianas a rather distinctive appearance. The species number about sixty and, with the exception of *Babiana socotrana* which is native in the island of Socotra off the coast of northeast Africa, they are confined to southern Africa.

The generic name *Babiana* is derived from the local Cape name 'Babianer', which in turn is a corruption of the Afrikaans name 'Bobbejane' for the baboons which enjoy eating the corms of these plants. Although the variety Redouté has illustrated here was described as a species in 1781 by Linnaeus the younger, he regarded it as a *Gladiolus* and it was not until 1802 that John Gawler (after 1804 John Bellenden Ker, but known to botanists as Ker-Gawler) created *Babiana* as a genus.

In the wild *Babiana tubulosa* variety *tubiflora* is a fairly frequent plant, growing in sandy places in coastal areas of the southwest Cape, from Clanwilliam to Riverside. None of the babianas can be said to be popular or well-known plants in European or American gardens, although they are quite showy greenhouse plants. The modern hybrids exist in a great range of colours which nearly equals that of the related and much-loved freesia.

1–2. Bracts 3. Flower 4. Bractiole 5. Pistil
6. Section of ovary 7. Flower opened out 8. Stamen

Gladiolus Tubiflorus

Tritonia crocata (Linnaeus) Ker-Gawler
figured as *Ixia crocata*

The tritonias were included in the genus *Ixia* until in 1802 John Bellenden Ker-Gawler established a separate genus for those species which had brown papery bracts subtending the flowers and stamens which seemed to point in various directions. This latter feature led him to liken them to a weathercock, the connection being that Triton, son of Poseidon, was supposed to be able to raise waves on calm waters by blowing on a shell. It is indeed a fact that the stamens of *Tritonia squalida*, and a few of the related species, are disposed in random directions, so that the name is appropriate for these. However, in most of the forty or so species in the genus they are more stable in orientation.

The genus is almost entirely South African and the species occur mainly in the southwest winter-rainfall region, although some are present as far north as Mozambique. In the past they were thought to be linked with *Ixia*, but Dr P. Goldblatt who has made a thorough study of the genera of South African Iridaceae considers that *Tritonia* is in fact more closely related to *Crocosmia*, the well-known Montbretia in our gardens.

Tritonias may be recognized by their flowers, which have nearly equal, broad overlapping perianth segments, produced in one-sided spikes. *Tritonia crocata* and *Tritonia squalida*, illustrated here, belong to a small group of very similar-looking species with orange, red, yellow or pinkish flowers. The latter, which was wrongly named *hyalina* in the original text of *Les Liliacées*, has pale pinkish or purplish flowers with a conspicuous transparent portion at the base of each of the perianth segments, while *Tritonia crocata* is usually a much more brilliant saffron colour. However, this species has received the attentions of horticulturists and many colour forms have been selected, ranging from white to yellow and bright orange or red. Some of these are almost certainly hybrids with other species.

Tritonias are easy to cultivate and last well as cut flowers.

1–2. Bracts   3. Flower opened out   4. Pistil

Ixia Crocata

IRIDACEAE

Tritonia squalida (Aiton) Ker-Gawler
figured as *Ixia hyalina*

1. Flower opened out   2. Pistil

Ixia Hyalina

IRIDACEAE

Sparaxis grandiflora (D. Delaroche) Ker-Gawler
figured as *Ixia grandiflora*

SPARAXIS

Sparaxis is a small genus of only six species, all of which occur in the southwestern Cape region of southern Africa. Before 1805 they were included with *Ixia*, but they were then segregated by John Bellenden Ker-Gawler because of the papery bracts subtending the flowers. These are often lacerated at the apex and the new generic name, taken from the Greek word *sparasein*, torn, refers to this feature. Later taxonomists however showed that Ker-Gawler's concept of *Sparaxis* was not entirely correct and that some of the species he included in his new genus belonged in fact to other genera which were subsequently called *Synnotia* and *Dierama*. The genus *Sparaxis*, as defined recently by P. Goldblatt in the *Journal of South African Botany*, Volume 35 (1969), includes *Streptanthera*, a similar genus described by R. Sweet in 1827, separated solely on the basis of having twisted anthers and larger style branches.

The two species of *Sparaxis* illustrated here are among the most colourful and are quite variable. *Sparaxis tricolor* is usually orange-red or sometimes paler orange, with darker red or blackish markings nearer to the base of each of the segments and a yellow tube giving the tri-colour effect. It occurs in the wild only near the town of Nieuwoudtville, where it flowers in September in dampish clay, which however becomes very dry during the southwest Cape summer. The species of *Sparaxis* hybridize and many garden forms have been given the name *Sparaxis tricolor* but are in fact not the true wild plant.

Sparaxis grandiflora is a much more widespread species and has a great range of variation, so much so that Goldblatt has distinguished four subspecies. In flower colour alone it may have a white, reddish-purple, violet or yellow ground with or without black blotches and darker streaking. The perianth tube may be purple, yellow or black. In leaf character, too, there is great variation, from the narrowly lanceolate to the broadly sickle-shaped. The plant illustrated by Redouté is a very dark violet form probably nearest to subspecies *grandiflora*.

1–2. Bracts 3. Flower opened out 4. Pistil

Ixia Grandiflora

IRIDACEAE

Sparaxis tricolor (Curtis) Ker-Gawler
figured as *Ixia tricolor*

1. Pistil 2. Corm

Ixia Tricolor

Tigridia pavonia (Linnaeus fil.) de Candolle

TIGER FLOWER

The gaudy Tiger Flower, or Tigridis Flos of the Spanish Conquistadores, was known long before the invasion of Mexico, under the Aztec name of 'Ocoloxochitl'. In addition to its obvious use as a decorative plant, its bulbs were cooked and eaten, in which form it was known as 'Cacomitl'. The name 'Cacomite' survives today and is used by Mexicans for *Tigridia* species in general. The plant was not noted by botanists until 1576 when Matthias de Lobel mentioned it in his *Plantarum Historia* and provided a recognizable woodcut illustration. The first name to be published correctly under the Rules of Nomenclature was in 1781 when Linnaeus the younger placed it in the genus *Ferraria*, a superficially similar but unrelated genus from South Africa. The name now accepted, in the genus *Tigridia*, was given by A.P. de Candolle in Redouté's *Les Liliacées* in 1802. The name refers to the spotted American 'tiger' or jaguar.

The long history of cultivation has led to some obscuring of the actual wild provenance of the plant and although generally accepted as a native of Mexico, it has been much dispersed by man and is now growing, often apparently wild, throughout Central America and in several South American countries. It is so easy to cultivate that it is well known as a showy garden plant in many countries, even in cool temperate climates where the bulbs must be lifted for the winter months.

It seems churlish to look for faults in such a striking plant, but one criticism that might be levelled at *Tigridia pavonia* is that the invidual flowers are so short-lived. They open in the early morning and shrivel at about mid-day and although several flowers are produced from each inflorescence one cannot be sure that there will be blooms each day. A large planting of bulbs will however ensure a display over a long period. Seeds are produced very prolifically and from these the bulbs will reach flowering size in only one or two years, more rapidly than almost any other bulbous plant.

Apart from the common reddish-orange form there are a number of other variants, having white, pink or yellow flowers, with or without blotches in the centre.

1. Bulb 2. Pistil and stamens 3. Capsule
4. Section of capsule showing seeds

Tigridia Pavonia

Witsenia maura Thunberg

The Waaiertje, to give this curious plant its local name, is one of the few members of the iris family to have woody stems. For this reason it is thought to be one of the more ancient elements present in the South African flora. *Witsenia maura* is the only species in the genus, but the equally interesting genera *Klattia* and *Nivenia* also have this shrubby type of growth. In *Witsenia* the stems can reach to almost 2 metres high and they carry fans of leaves which are crowned by the heads of blue and yellow long-tubed flowers, each of which is about 5 centimetres long. Surrounding the flowers is an involucre of tough, overlapping shiny yellowish bracts. The blue colour of the flowers varies to some extent and in some forms can appear almost black, while on the exterior of the perianth segments is a yellowish fur of hairs. The flowers are scented and pollination is carried out by the Sunbirds and Sugarbirds.

Witsenia is a rare plant in the wild and grows in swampy places in the southwest Cape region. It was first found by C.P. Thunberg at Noordhoek and False Bay on the Cape Peninsula. The genus was described by him in 1782 and is named after Nicolaus Witsen, a Dutch burgomaster who cultivated South African plants in the eighteenth century.

1. Flowers and bracts 2. Outer bract 3. Inner bracts and flowers 4. Flower 5. Perianth 6. Stamen (enlarged) 7. Pistil

Witsenia Maura

MUSACEAE

Musa uranoscopus Colla
figured as *Musa coccinea*
WILD BANANA

This is one of the wild bananas, a native of southeast Asia and sometimes cultivated as an ornamental plant for its bright red bracts. The fruits are not as fleshy as in the well-known *Musa paradisiaca* which is illustrated on page 233 and are not used as a food crop in the same way.

Musa and another genus, *Ensete*, together form the family Musaceae. They are large herbs—not trees as is sometimes believed—the stems of which are made up of the tough leaf stalks or sheaths which are rolled tightly around each other. The huge leaf blades are also rolled up and when they unfold they are entire, without any lacerations on the margin. It is the wind and heavy rain of tropical storms which give the leaves their characteristic tattered appearance, since the veins are parallel and have no lateral strength to hold the leaf together. These parallel fibres are however quite tough and some species of *Musa* are used locally as a fibre crop. *Musa textilis*, for example, is the source of Manila hemp.

The flowers are produced in racemes, upright or pendulous, and are enclosed in the bud stage by brightly coloured bracts. These flowers are rich in honey and attract bees and small birds.

In *Musa uranoscopus* the inflorescence is erect and the tightly sheathing bracts are red with a yellow or green apex, while the flowers are yellow or orange with greenish tips. This is an extremely colourful species and not surprisingly is much used for ornamental purposes in congenial climates. Even in Britain its potential as an attractive greenhouse plant was realized at an early date and it is said to have been cultivated in Stepney by a Mr Evans in 1792.

Musa Coccinea

MUSACEAE

Musa paradisiaca Linnaeus

BANANA

There are probably over two hundred different forms of the banana in cultivation around the world and it is impossible to say which of these is illustrated here by Redouté. *Musa sapientum*, or *paradisiaca* subspecies *sapientum*, is the banana which is most commonly eaten raw, while *Musa paradisiaca* is the plant whose fruits are higher in starch and less sugary and are much used in parts of the tropics for cooking. These are known as Plantains and are steamed or roasted, or dried and ground into flour. Other varieties of banana are used for making alcoholic drinks, especially in parts of South America and East Africa.

Countries that are unable to produce their own bananas import large quantities from the tropics; thus only the few varieties which travel well are known to them, and even these are not distinguished by varietal names in the shops. There are many with even better-flavoured fruits, but because these do not keep well during the period of transport they are unknown to most Europeans. Some have short fruits and are known as Lady's Fingers, while others have orange-coloured flesh, or reddish skins. In some subtropical or warm temperate countries such as the Canaries, Israel and southern Turkey it is possible to grow a dwarf variety, known as *cavendishii*, which has smaller fruits.

These cultivated forms of *Musa*, grown for their succulent fruits, are not true wild species and are mostly sterile. They are propagated by suckers which arise from the rhizome, the new plant forming a flowering stem within a year of planting. The banana is thus very easily grown and yields more food per acre than even the potato. In a successful plant, each inflorescence may give rise to a bunch of as many as two hundred individual bananas; these are produced in clearly defined groups on the axis; it is in the form of these groups, or 'hands' as they are called, that bananas are usually sold on the retail market.

The wild species of *Musa*, of which there are about thirty-five, are distributed through the Old World tropics. They have smaller fruits containing large seeds and are therefore much less attractive as food plants.

Musa Paradisiaca

HELICONIACEAE

Heliconia humilis Jacquin

LOBSTER CLAW

This banana-like plant is placed by some authorities together with *Musa* in the family Musaceae, but it is now generally accepted that it is best placed in Heliconiaceae. It is easy enough to distinguish between the true banana (*Musa*) and the heliconia, since the leaves of the latter, and the bracts of the flowering stems, are two-ranked, whereas in *Musa* they are spirally arranged; that is to say they are produced all round the stem rather than on opposite sides of it. Also, *Heliconia* species have capsules instead of fleshy banana-like fruits. In addition there is a distinct geographical separation between the two groups, *Musa* species being native to the Old World tropics and *Heliconia*, about one hundred and twenty species in all, being primarily tropical American with a few extending to the western Pacific region.

Heliconia humilis, commonly known as the Lobster Claw, is one of the smaller species in the genus; hence its specific epithet, meaning 'low-growing'. Even so it is far from being a dwarf plant, for the stem can reach a height of 1.2 metres and the leaves are up to 1.5 metres long and 28 centimetres wide. Like several of the other species, it is a striking plant suitable for cultivation in tropical gardens or in warm greenhouses and conservatories of the temperate zones. It has been grown in European gardens for well over a hundred years and was recorded at Kew Gardens in 1866.

Heliconia humilis is a native of Guyana, Venezuela and Caracas, where it grows in marshy places.

1. Bract 2. Flower 3. Pistil and stamens 4. Pistil
5. Young fruit 6. Section of fruit 7. Mature fruit 8. Seed

Heliconia Humilis

STRELITZIACEAE

Strelitzia reginae Aiton

BIRD OF PARADISE FLOWER

This extraordinary plant caused a great sensation in Britain when it flowered at Kew Gardens. Indeed, in 1790 the compilers of the *Botanical Magazine* took the unprecedented step of providing a folding plate to accommodate its size, 'in order that we may give our readers an opportunity of seeing a coloured representation of one of the most scarce and magnificent plants introduced into this country.' Even before this date Sir Joseph Banks had seen the plant and named it *Strelitzia* in honour of Charlotte of Mecklenburg-Strelitz who became Queen to George III in 1761.

Strelitzia rapidly became a popular, although expensive, greenhouse plant and it is noted that in the late eighteenth century plants were available from South Africa at three guineas each, a considerable sum.

Together with two, small, less well-known genera, *Strelitzia* forms the family Strelitziaceae. There are five species in the genus, all rather banana-like herbs with a robust rhizomatous rootstock and long-stalked leaves. The whole inflorescence is enclosed within the large tubular beak-like bract and it is doubtless this feature which inspired the alternative vernacular name of Crane Flower. All the species are native to the south and southeast Cape region of South Africa and all are equally impressive, some of them having striking pure white flowers in deep purple-blue bracts.

The highly developed flowers require a little explanation as regards their structure, although they follow the general pattern of floral morphology in a monocotyledon by having six perianth parts. The outer three segments are orange-yellow and more or less equal, while the inner three blue ones are very unequal. Two of these latter are mirror images of each other and are held together to form an arrow-shaped structure, while the third is much smaller and somewhat boat-shaped. A succession of these blooms is produced from the large sheathing spathe or bract.

1. Inflorescence 2. Flower with bracts removed
3. Flower with perianth segments removed
4. Pistil and stamens 5. A stamen 6. A bract

Strelitzia Reginæ

Alpinia zerumbet (Persoon) B.L. Burtt & R.M. Smith figured as *Globba nutans*

SHELL FLOWER, PORCELAIN LILY, SHELL GINGER

This plant has the largest and showiest flowers in the genus *Alpinia*, which constitutes a group of about two hundred and fifty species distributed through the warmer parts of Asia and Polynesia. The name *Alpinia* commemorates Prospero Alpino (1553–1616), an Italian botanist and professor of botany at Padua, who studied the plants of Egypt in particular.

This species has long been known in cultivation and is said to have been brought into Europe in 1792. Not surprisingly it received the attentions of many botanists and artists and there was much confusion over its name (see the account of B.L. Burtt and R.M. Smith in *Notes from the Royal Botanic Garden Edinburgh*, Volume 31, 1972). Thus the species may be found under various names such as *Globba nutans, Alpinia nutans, Alpinia speciosa* and *Zerumbet speciosum*, and at times has even been included in the genera *Languas* and *Catimbium*.

Alpinia zerumbet is very widely cultivated in the tropics outside its native area, which is thought to be northeast India, Burma, Indo-China and possibly northern Malaysia. Its gorgeous sweetly scented orchid-like flowers are produced throughout most of the year and it is at its best when growing near water, reaching a height of 3 to 6 metres. In Delhi it is regarded as one of the loveliest of all ornamental plants and is often grown in parks and gardens. The thick waxy texture and glistening appearance of the flowers has suggested the local names Porcelain Lily and Shell Ginger, the latter because of the plant's membership of the family Zingiberaceae to which the true ginger, *Zingiber officinale*, belongs.

Apart from its quite outstanding ornamental value, *Alpinia zerumbet* has also provided and perhaps still does on a small local scale, a substitute for 'galangal', a substance obtained from the roots of *Alpinia officinarum* and used in the treatment of respiratory complaints and for rheumatism. Also the fibrous stems have been utilized for making paper.

Outside the tropics, *Alpinia zerumbet* can be grown in greenhouses, although it is mainly confined to botanic gardens on account of its vigorous growth.

1. Flower 2. Flower with bracts removed
3. Section of flower showing pistil and stamen

Globba Nutans

INDEX OF FLOWERS ILLUSTRATED